NELSON QMATHS

ESSENTIAL MATHEMATICS

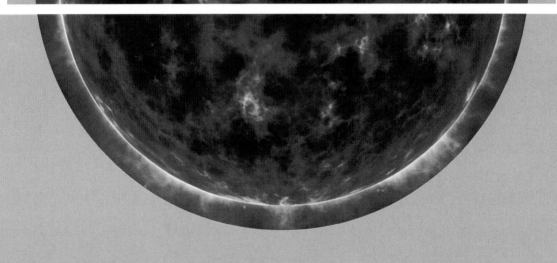

YEAR **11**

Sue Thomson
Judy Binns

T0362726

Nelson QMaths 11 Essential Mathematics
1st Edition
Sue Thomson
Judy Binns

Publishing editor: Robert Yen, Alan Stewart
Project editor: Alan Stewart
Editor: Anna Pang
Cover design: Chris Starr (MakeWork)
Text design: Nicole Melbourne
Project design: Petrina Griffin
Cover image: iStock.com/alexandarilich
Permissions researcher: Kaitlin Jordan
Production controller: Christine Fotis
Typeset by: Cenveo Publisher Services

Any URLs contained in this publication were checked for currency during the production process. Note, however, that the publisher cannot vouch for the ongoing currency of URLs.

For product information and technology assistance,
in Australia call **1300 790 853**;
in New Zealand call **0800 449 725**

For permission to use material from this text or product, please email
aust.permissions@cengage.com

National Library of Australia Cataloguing-in-Publication Data
Thomson, Sue, 1958- author.

Nelson Qmaths11: essential mathematics / Sue Thomson, Judy Binns.

9780170412650 (paperback)
For secondary school age.

Mathematics--Study and teaching (Secondary)
Mathematics--Textbooks.

Other Creators/Contributors:
Binns, Judy, author.

Cengage Learning Australia
Level 7, 80 Dorcas Street
South Melbourne, Victoria Australia 3205

Cengage Learning New Zealand
Unit 4B Rosedale Office Park
331 Rosedale Road, Albany, North Shore 0632, NZ

For learning solutions, visit **cengage.com.au**

Printed in China by China Translation & Printing Services.

CONTENTS

UNIT 1

1

2

3

UNIT 2

PREFACE

Nelson QMaths, Queensland's longest-running senior mathematics series, has been rewritten for the new syllabuses and assessment procedures for implementation from 2019. Based on the Australian Curriculum, four new senior mathematics courses have been introduced into Queensland schools.

- Essential Mathematics
- General Mathematics
- Mathematical Methods
- Specialist Mathematics

With the introduction of new assessment procedures, *Nelson QMaths* will have a renewed focus on assessment, and include features such as chapter reviews, practice sets (mixed reviews) and video tutorials. In this book, teachers will find familiar features such as clear worked examples, graded exercises, strong syllabus coverage, Investigation, Technology and a glossary/index. We wish all teachers and students using this book every success in embracing the new mathematics courses.

ABOUT THE AUTHORS

Sue Thomson is an experienced teacher and educational leader. She was an examination writer, assessor, marker and curriculum writer, and a councillor for the Australian Association of Mathematics Teachers. Sue is a prolific and successful author with an interest in language development, financial literacy and making mathematics accessible to all, especially senior practical mathematics and thematic mathematics. With Ian Forster, she wrote the successful *Access to Prevocational Maths* series.

Sue dedicates this book to the memory of her husband and co-author, **Ian Forster**.

Judy Binns is a mathematics coordinator and experienced author who has taught in urban and rural schools. She has an interest in motivating students with learning difficulties, and wide experience in teaching senior practical mathematics courses. Judy often presents at local and state conferences.

More recently, Sue and Judy wrote *New Century Maths 11-12 Mathematics Standard 1* for NSW, *Nelson Senior Maths 11-12 Essentials* for WA and the ACT, and *Nelson VCE Foundation Mathematics* for Victoria.

CONTRIBUTING AUTHORS

Deborah Van Hoek wrote many of the *NelsonNet* worksheets.

John Drake, Katie Jackson and **Joanne Magner** created the video tutorials.

SYLLABUS REFERENCE GRID

Topics and subtopics	Nelson QMaths 11 Essential Mathematics chapter
UNIT 1: NUMBER, DATA AND GRAPHS	
Number	
Ratios	6 Colourful ratios
Rates	4 Applying rates
Percentages	2 Giving 110%
	7 Applying percentages
Representing data	
Classifying data	8 Show me the data
Data presentation and interpretation	5 Show me the graph
	8 Show me the data
Graphs	
Reading and interpreting graphs	5 Show me the graph
Drawing graphs	5 Show me the graph
Using graphs	9 Practical graphs
UNIT 2: MONEY, TRAVEL AND DATA	
Managing money	
Earning money	10 Earning money
	14 Paying tax
Budgeting	14 Paying tax
Time and motion	
Time	11 It's about time
Distance	16 Going places
Speed	13 Healthy figures
	16 Going places
Data collection	
Census	12 Census and surveys
Surveys	12 Census and surveys
Simple survey procedure	12 Census and surveys
Sources of bias	15 That's biased

Note: The Fundamental topic, Calculations, is first covered in Chapter 1 What's the score?, then integrated in all other chapters. Chapter 3 is a special **Problem-solving** chapter.

 NELSON QMATHS 11. Essential Mathematics ISBN 9780170412650

ABOUT THIS B⬤⬤K

AT THE BEGINNING OF EACH CHAPTER

- Each chapter begins on a double-page spread showing a **Chapter Problem** to be solved, a chapter table of contents, a **What we will do in this chapter?** list of outcomes, and a **How are we ever going to use this?** list of applications.

IN EACH CHAPTER

- Worked examples are explained clearly step-by-step, with the mathematical working shown on the right-hand side.

- Important facts and formulas are highlighted in a shaded box.

- Important words and phrases are printed in red and listed in the glossary at the back of the book.

- Graded exercises include **Problem solving** questions 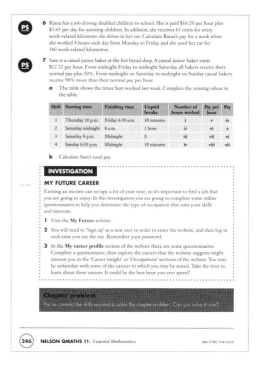, are linked to the worked examples and include exam-style problems and realistic applications.

- **Investigations** and **practical activities** explore the syllabus in more detail, providing ideas for modelling activities and assessment tasks.

6 Raina has a job driving disabled children to school. She is paid $16.20 per hour plus $3.65 per day for assisting children. In addition, she receives 65 cents for every work-related kilometre she drives in her car. Calculate Raina's pay for a week when she worked 4 hours each day from Monday to Friday and she used her car for 360 work-related kilometres.

7 Sam is a casual junior baker at the hot bread shop. A casual junior baker earns $12.32 per hour. From midnight Friday to midnight Saturday all bakers receive their normal pay plus 50%. From midnight on Saturday to midnight on Sunday casual bakers receive 98% more than their normal pay per hour.

a The table shows the times Sam worked last week. Complete the missing values in the table.

Shift	Starting time	Finishing time	Unpaid breaks	Number of hours worked	Pay per hour	Pay
1	Thursday 10 p.m.	Friday 6:30 a.m.	30 minutes	i	v	ix
2	Saturday midnight	8 a.m.	1 hour	ii	vi	x
3	Saturday 8 p.m.	Midnight	0	iii	vii	xi
4	Sunday 6:30 p.m.	Midnight	30 minutes	iv	viii	xii

b Calculate Sam's total pay.

INVESTIGATION

MY FUTURE CAREER

Earning an income can occupy a lot of your time, so it's important to find a job that you are going to enjoy. In this investigation you are going to complete some online questionnaires to help you determine the type of occupation that suits your skills and interests.

1 Visit the **My Future** website.

2 You will need to 'Sign up' as a new user in order to enter the website, and then log in each time you use the site. Remember your password.

3 In the **My career profile** section of the website there are some questionnaires. Complete a questionnaire, then explore the careers that the website suggests might interest you in the 'Career insight' or 'Occupations' sections of the website. You may be unfamiliar with some of the careers to which you may be suited. Take the time to learn about these careers. It could be the best hour you ever spend!

Chapter problem

You've covered the skills required to solve the chapter problem. Can you solve it now?

10.03 Bonuses and allowances

Some jobs include **allowances** for doing unpleasant work, for working under difficult conditions, or to cover expenses such as uniform and travel.

Some jobs pay **bonuses** (extra pay) for doing good work, meeting targets or deadlines.

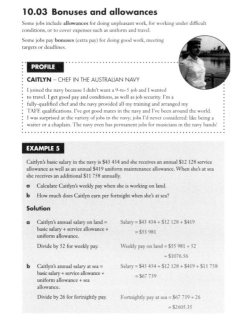

PROFILE

CAITLYN – CHEF IN THE AUSTRALIAN NAVY

I joined the navy because I didn't want a 9-to-5 job and I wanted to travel. I get good pay and conditions, as well as job security. I'm a fully-qualified chef and the navy provided all my training and arranged my TAFE qualifications. I've got good mates in the navy and I've been around the world. I was surprised at the variety of jobs in the navy; jobs I'd never considered: like being a waiter or a chaplain. The navy even has permanent jobs for musicians in the navy bands!

EXAMPLE 5

Caitlyn's basic salary in the navy is $43 434 and she receives an annual $12 128 service allowance as well as an annual $419 uniform maintenance allowance. When she's at sea she receives an additional $11 758 annually.

a Calculate Caitlyn's weekly pay when she is working on land.

b How much does Caitlyn earn per fortnight when she's at sea?

Solution

a Caitlyn's annual salary on land = basic salary + service allowance + uniform allowance.

Salary = $43 434 + $12 128 + $419
= $55 981

Divide by 52 for weekly pay.

Weekly pay on land = $55 981 ÷ 52
= $1076.56

b Caitlyn's annual salary at sea = basic salary + service allowance + uniform allowance + sea allowance.

Salary = $43 434 + $12 128 + $419 + $11 758
= $67 739

Divide by 26 for fortnightly pay.

Fortnightly pay at sea = $67 739 ÷ 26
= $2605.35

EXAMPLE 6

Sonia is paid $15.48 per hour as a security guard. Each week she receives an additional $61.05 for her guard dog and $6.75 for her torch. Sonia receives $14.15 per shift travel allowance.

Sonia works a 4-hour shift, 6 nights per week. How much is she paid per week?

Solution

Sonia's total weekly pay

= wages + allowances + dog + torch

Wages = 4 × 6 × $15.48
= $371.52

Travel allowance = 6 × $14.15
= $84.90

Total weekly pay = $371.52 + $84.90 + $61.05 + $6.75
= $524.22

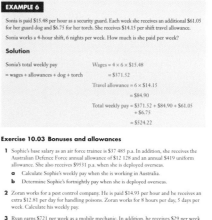

Exercise 10.03 Bonuses and allowances

1 Sophie's base salary as an air force trainee is $37 485 p.a. In addition, she receives the Australian Defence Force annual allowance of $12 128 and an annual $419 uniform allowance. She also receives $9531 p.a. when she is deployed overseas.

a Calculate Sophie's weekly pay when she is working in Australia.

b Determine Sophie's fortnightly pay when she is deployed overseas.

2 Zoran works for a pest control company. He is paid $14.93 per hour and he receives an extra $12.81 per day for handling poisons. Zoran works for 8 hours per day, 5 days per week. Calculate his weekly pay.

3 Ryan earns $721 per week as a mobile mechanic. In addition, he receives $29 per week for work-related phone calls and $0.60 per kilometre for work-related travel. Calculate Ryan's pay for a week in which he drove 420 km in his truck for work.

4 Kate is the manager of a fast food chain. She is paid $28 per hour for a 35-hour week plus a $8.30 per week laundry allowance. She receives a $30 bonus for every accident-free week at the shop and another $95 bonus if the shop makes $100 000 or more in sales. Last week the shop was accident-free and the sales were $110 000. How much was Kate paid last week?

5 Zack drives a furniture removal truck. He is paid $15.12 per hour Monday to Friday, time-and-a-quarter on Saturday and double time on Sunday. He receives a flat fee of $12.59 per day for handling heavy furniture. Calculate Zack's pay for a week when he delivered heavy furniture for 33 hours Monday to Friday, 6 hours on Saturday and 3 hours on Sunday.

AT THE END OF EACH CHAPTER

- **Keyword activity** focuses on the mathematical language and terminology learned in the chapter.

- **Solution to the chapter problem** revisits the problem introduced at the start of the chapter and solves the problem using 4 stages: WHAT?, SOLVE, CHECK and PRESENT.

- **Chapter review** contains revision exercises that include **Problem solving** and are linked to chapter exercises.

- **Practice sets** revise the skills and knowledge of previous chapters.

AT THE END OF THE BOOK

- **Glossary/Index** is a comprehensive dictionary of course terminology.

- **Answers**.

NELSONNET TEACHER WEBSITE

Margin icons link to print (PDF) and multimedia resources found on the NelsonNet teacher website, **www.nelsonnet.com.au**. These include:

- **Worksheets** and **puzzle sheets** that are write-in enabled PDFs

- **Skillsheets** of examples and exercises of prerequisite skills and knowledge

- **Video tutorials**: worked examples explained by 'flipped classroom' teachers

- **Spreadsheets**: *Excel* files

- **Weblinks**

- A **teaching plan**, in Microsoft Word and PDF formats

- **Chapter PDFs** of the textbook

- **Resource Finder**: search engine for *NelsonNet* resources

Note: Complimentary access to *NelsonNet* is only available to teachers who use this book as a core educational resource in their classroom. Contact your sales representative for information about access codes and conditions.

NELSON QMATHS 11–12 SERIES

CALCULATIONS

1.

WHAT'S THE SCORE?

Chapter problem

Ben bought fish and chips for $11.25 and paid with a $50 note.
The cashier gave him a $20 and a $5 note and a few coins for his change.
Could the change be wrong?

WHAT WILL WE DO IN THIS CHAPTER?

- Calculate with numbers, including negative numbers and decimals
- Add, subtract, multiply and divide using order of operations, with and without a calculator
- Estimate and check the reasonableness of answers
- Understand place value after the decimal point in decimals
- Round numbers to decimal places
- Solve practical problems involving numbers

HOW ARE WE EVER GOING TO USE THIS?

- Keeping score in sporting events
- Shopping
- At work
- Budgeting: for yourself or for a party
- Every day

1.01 Target practice

Alan coaches a junior football team. He designed a ball-throwing competition to help his players improve the accuracy of their passes. Each player throws a football at the target two times each turn. The first player to reach 100 points is the winner.

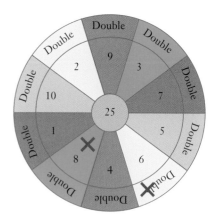

The 2 red crosses on the diagram show where Marc's throws landed. One throw landed in **double 6** and the other in **8**.

Marc's score = 2 × 6 + 8

$\qquad$ = 12 + 8

$\qquad$ = 20

Exercise 1.01 Target practice

1 Without using a calculator, work out the scores for each turn.

a
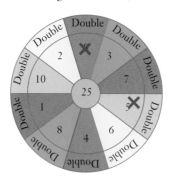

b
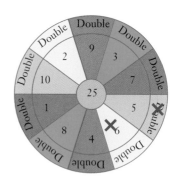

c
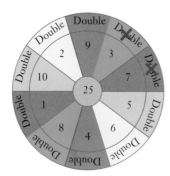

d
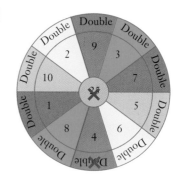

2 Add each set of scores as quickly as you can.

   a A 7 and a 9   **b** Double 5 and 8   **c** Double 6 and double 4

   d Double 9 and double 5   **e** Bullseye and double 10   **f** A 3 and double 7

3 Jason's score from 2 throws was 13. Suggest 5 different combinations of points that can make a score of 13.

4 How many different ways can you get a score of 7 from 2 throws? List the ways.

5 To win the game, a player's score must add to exactly 100. David's score is 77.

    a How many more points does David need to score to win?

    b What strategy can David use to win on his next turn?

6 What is the smallest number of throws required to make a total of exactly 100? How is it achieved?

7 Courtney's score is 81. She decided to aim for a 9 and a 10. With her first throw, Courtney missed the 9 and her ball landed on 3. How can she still make 100 on her second throw?

8 When Alan designed the positions of the numbers on the target, he didn't put any big numbers next to each other. He put small numbers on each side of the big numbers. Why do you think Alan arranged the numbers in this way?

Imagefolk / One Shot/PNZ / Arno Gasteiger

1.02 Playing darts

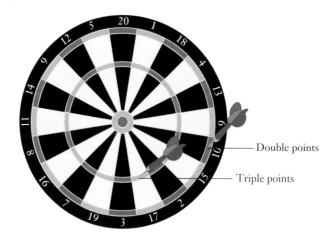

Double points

Triple points

There are 20 numbers on a dartboard and each player throws 3 darts on their turn. If a dart lands in the **outside ring** of a number, the dart scores **double points**. It scores **triple points** if it lands in the **inside ring**. The **bullseye** is worth 50 points. The green ring around the bullseye is worth 25 points.

Exercise 1.02 Playing darts

1 Calculate the score for each player.

a

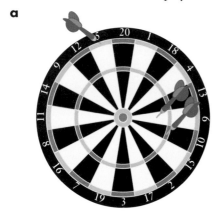

b

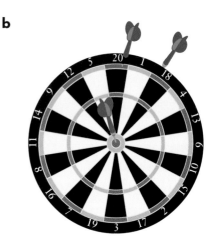

2 What is the highest score possible with 3 darts?

3 How can you score a total of 37 using 3 darts?
Suggest 2 possible ways.

4 In the dart game '201', players start with 201 points and they keep subtracting the points they score with 3 darts. The winner is the first to reach 0, but they must finish with a double. If a player's score on one turn is more than their remaining points, *no* points are subtracted for that turn. Use the rules of the game to answer the rest of the questions in this exercise.

With her first set of 3 darts, Madeleine scored a 7, a double 5 and a triple 11.

How many points did Madeleine have left after this?

5 After the second round, Tim had 152 points left and Rebecca had 147. Who is winning the game: Tim or Rebecca?

6 Fiona had 126 points remaining before her turn. After her turn, she had 103 points. One of her darts landed on double 1 and another on 5. Where could her third dart have landed?

7 Divya's score was 86. She threw 2 double 20s and a triple 19. What was Divya's score after her turn?

8 Before he had his turn Santo had 136 points remaining. His first dart landed on 20, his second landed on triple 4 and his third missed the board. How many points did Santo have left after his turn?

9 Andre needs to score 34 points, finishing with a double, to win a game of 201.

How could he win using:

a   one dart? **b**   2 darts? **c**   3 darts?

10 What is the largest number of points a player can have and still win on their next turn? Explain your answer.

11 Renata and Samantha are playing 201, writing their progress scores on a board.

Turn	Renata	Samantha
	201	201
1	162	175
2	102	143
3	63	85
4	39	55
5	24	29
6	0	

a   In what number turn did Samantha score a total of 30 points?

b   Who shot 3 darts in the single number 20 in the same turn? In which turn did she do it?

c   What was Samantha's biggest score in one turn?

d   In which turn did Samantha throw a double 3, triple 2 and single 20?

e   Renata finished when she threw a double 7. Explain how it was possible for her to finish with a double 7.

f   Samantha can tie the game if she can score a total of 29, finishing with a double. Explain why Samantha will need more than one throw to score 29.

g   What strategy do you suggest Samantha should adopt to try to tie the game?

1.03 Negative scores

Integers are positive and negative whole numbers and zero.

Scores in golf and indoor cricket both make use of **negative numbers**. In golf, a negative number is a good score, but in indoor cricket it's a bad score.

> Golf has a language of its own.
> * A 'birdie' is 1 stroke under par, or –1.
> * An 'eagle' is 2 strokes under par, or –2.
> * An 'albatross' is 3 strokes under par, or –3.
> * A 'bogie' is 1 stroke over par, or +1.
> * A 'double-bogie' is 2 strokes over par, or +2.

Each hit of a golf ball is called a **stroke**. The course rating is called 'par', meaning average score. If a golfer's score is –2, it means that she took 2 fewer strokes than the course rating to complete her game. If an indoor cricketer's score is –2, it means that he lost 2 more runs than he scored.

EXAMPLE 1

Jake and Megan played very well in a golf championship. Jake finished with a score of –2 and Megan's score was –3.

a What do scores of –2 and –3 mean?

b Did Jake or Megan play the better game? Explain your answer.

Solution

a In golf, negative scores are good scores. Every 1 negative number means 1 better than the course rating.

–2 means 2 strokes better than the course rating, and –3 means 3 strokes better.

b In golf, –3 is a better score than –2.

Megan played the better game.

Exercise 1.03 Negative scores

1 Angelo and Brett are playing golf. Angelo's score is +2 and Brett's score is –1. Who is winning the game?

2 Last week, Darren's golf score was +3. Today, his score was 4 strokes better. What was Darren's score today?

3 This sign shows the leaderboard at the end of a golf tournament. Par, or the course rating, is 72 strokes.

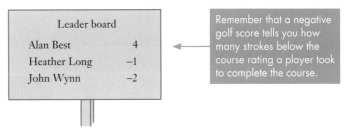

Remember that a negative golf score tells you how many strokes below the course rating a player took to complete the course.

Leader board

Alan Best	4
Heather Long	−1
John Wynn	−2

  a How many strokes did John take to complete the course?

  b How many more strokes did Alan take than Heather?

4 Shari is going snow skiing.

  a What does a temperature of −10°C mean?

  b Which temperature is the coldest: 5°C, −4°C or 0°C?

  c The temperature on the ski fields today is −3°C. Tomorrow's forecast temperature will be 4°C colder. What temperature is forecast for tomorrow?

  d At lunchtime the temperature was −7°C. The temperature went up 2°C between lunch and afternoon tea. What was the temperature at afternoon tea time?

5 In indoor cricket, batters score runs by hitting the ball into specially-marked zones and by running between the batting and running creases. Every time a batter is given out, 5 runs are deducted from the score, but the batter continues with his or her innings.

  a Jesinta scored 8 runs and was given out twice. What was her final score?

  b Shane scored 46 runs but his final score was only 31. How many times was he given out?

  c Grant and Mark are batting together. Here are their statistics.

Batter	Zone A 1 run hits	Zone B 2 run hits	Zone C 4 run hits	Zone D 6 run hits	Runs scored by running	Number of outs
Mark	4	8	2	1	12	3
Grant	2	4	6	3	21	2

Together, Mark and Grant scored $\frac{1}{4}$ of their team's runs. How many runs did the team score?

  d The batting team is 8 runs in front and there are 4 balls remaining. How is it possible for the batting team to lose?

NEGATIVE ENVIRONMENTS!

Lake Eyre in South Australia is −16 m above sea level. What does this mean?

What you have to do

- Investigate the location of Lake Eyre. What is the relevance of the negative sign in '−16 m'?

- Find other locations around the world that have a negative height above sea level.

- In the photograph of a sign from a Dutch department store, one of the floors is labelled −1. What does this mean?

- What does a bank mean when it says 'The amount owing on this credit card is −$30'?

- Find examples of negative numbers in banking. You could investigate credit card statements, loan statements or internet banking sites.

- Explain how banks use negative numbers.

Photo courtesy Sue Thomson

Order of operations

1.04 Order of operations

What is the answer to $200 \div 20 \times 2$? Is it 20 or 50?

BIDMAS is an easy way to remember the **order of operations** in a **mixed calculation**.

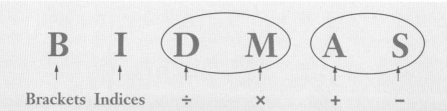

To solve this problem, we have special rules for the order in which we do calculations.

- **B**rackets ( ) first
- **I**ndices (powers) next
- Then any **D**ividing ÷ or **M**ultiplying ×, working from left to right
- Finally, any **A**dding + or **S**ubtracting −, working from left to right

So $200 \div 20 \times 2 = 20$, because we do × and ÷ at the same time and we work from left to right.

EXAMPLE 2

Evaluate each mixed expression.

a $60 - 2 \times 5^2$

b $(12 + 5 \times 3) \div 9 + 4$

c $11 + (5 + 11) \div (12 \div 3)$

Most calculators 'know' the order of operations. If you press the [=] key only at the end of the calculation, your calculator's answer will be correct. Check the answers in this example using a calculator.

Order of operations

Solution

a There are no brackets. Indices come next. $60 - 2 \times 5^2 = 60 - 2 \times 25$

The 5 is squared, so we do this first. $= 60 - 50$

Next comes multiply. $= 10$

Subtraction comes last.

b Do $\times$ before the $+$ inside the brackets first. $(12 + 5 \times 3) \div 9 + 4 = (12 + 15) \div 9 + 4$

Then do the $\div$. $= 27 \div 9 + 4$

The $+ 4$ comes last. $= 3 + 4$

$= 7$

c Do both sets of brackets first. $11 + (5 + 11) \div (12 \div 3) = 11 + 16 \div 4$

Then do the $\div$. $= 11 + 4$

The $+$ outside the brackets comes last. $= 15$

Exercise 1.04 Order of operations

1 Is each statement true or false? Write the correct value for any false statements.

Example **2**

a $4 + 2 \times 3 = 18$ **b** $12 - 3 \times 4 = 0$ **c** $20 \div 2 + 2 = 12$

d $6 + 18 \div 3 = 12$ **e** $2 \times 4^2 = 64$ **f** $48 \div 4 \times 3 = 4$

g $20 - 5 + 8 = 7$ **h** $5 \times (20 - 3 \times 4) = 40$ **i** $2 \times 5^3 = 250$

2 Find the value of each expression. Check your answers using a calculator.

a $18 - 3 \times 5$ **b** $24 \div (5 + 3)$ **c** $8 \times 3 - 10 \div 5$

d $(2 + 10) \times (12 - 9)$ **e** $3 \times (1 + 4)^2$ **f** $36 \div 12 \div 3$

g $30 \div 5 \times 2$ **h** $40 \times 2 \div 8$ **i** $4 \times (7 - 2) \div (3^2 + 1)$

j $4^2 + 5^2 - 3 \times 9$ **k** $10 + 5 \times 6$ **l** $300 - 20 \times 8$

m $120 \div 4 \times 5$ **n** $10 + 4^2$ **o** $5 \times (12 - 3 \times 2)$

3 Copy each statement and insert brackets to make the statement true.

a $4 + 7 \times 5 = 55$ **b** $60 \div 5 + 7 = 5$ **c** $3 \times 2^2 = 36$

d $6 + 8 \times 9 - 5 = 56$ **e** $3 \times 4 + 5 \times 2 = 34$ **f** $28 - 4 \times 5 \times 2 = 16$

4 When Siobhan used her calculator to evaluate $3 + 6 \times 5$, she got the wrong answer. She pressed the following calculator keys:

3 **+** 6 **=** **×** 5 **=**

Explain why Siobhan's answer was wrong.

5 In the game of **snooker** the points scored are determined by the colours of the balls sunk in the correct order. This table shows the value of each colour.

Colour	Red	Yellow	Green	Brown	Blue	Pink	Black
Points	1	2	3	4	5	6	7

In snooker, Brad sank 4 red balls, 2 brown balls, one pink and 2 blacks.
In the same game, he lost 15 points for foul shots.

a What does the expression $4 \times 1 + 2 \times 4 + 6 + 2 \times 7$ represent?

b Determine Brad's score for the game.

6 Mr Healy, the school principal, has a parent complaining about the marking of his daughter's maths exam. He claims his daughter's correct answer was marked wrong.

This is the question: $48 - 8 \times 3$

The daughter's answer was 120.

a Why was the daughter's answer wrong?

b How could Mr Healy explain why the daughter's answer is wrong?

c Put brackets in $48 - 8 \times 3$ to make the daughter's answer correct.

1.5 How much do I pay?

We put numbers to work in every aspect of our lives: from earning and spending money, through organising schedules, preparing food, assisting with our leisure activities to scoring sporting events.

EXAMPLE 3

Muspha's rent is $42 768 annually. How much is his rent per month?

Solution

← Annually means 'per year'.

There are 12 months in a year. Divide the annual amount by 12 to calculate the monthly amount.

Monthly rent = 42 768 ÷ 12
$$= 3564$$

Write your answer.

Each month Muspha pays $3564.

NELSON QMATHS 11. Essential Mathematics   ISBN 9780170412650

EXAMPLE 4

Gillian bought 3 books online and paid with her debit card. She paid $6.40 for postage and $7.95 for each book. Before Gillian bought the book the balance of her debit card was $160. Calculate the balance after her online purchase.

Solution

Calculate the cost of the postage and the books. Remember to do × before +.	Cost = $6.40 + 3 × $7.95
	= $30.25
	The books and postage cost $30.25.
Subtract the amount that Gillian spent from the balance of her debit card.	$160 − $30.25 = $129.75
Write your answer.	The balance of Gillian's debit card was $129.75.

Exercise 1.05 How much do I pay?

1 Finn's annual business expenses total $82 440. Calculate his average monthly expenses.

2 Gazi pays $3250 per month for rent. Calculate the annual rent that Gazi pays.

3 Li pays $12 per hour for parking. How much does he pay when he parks for 7 hours?

4 Aisling's business spends $1840 per fortnight on electricity.
   a How much does it spend per week on electricity?
   b Calculate the amount the business spends annually on electricity.

> Remember! A fortnight is 2 weeks long and there are 26 fortnights in a year.

5 Rob ordered 2 burgers and some fries for his lunch. The burgers cost $4.80 each and the fries cost $3.20. Rob paid with a $20 note. How much change should he get?

6 Voula spends 45 minutes at the gym 6 days per week. How many hours does she spend at the gym each week?

7 Joel plays golf 3 times per week. On average his golf games take 3 hours and 15 minutes.
   a How many hours does Joel play golf per week?
   b During holidays Joel plays golf every day. Calculate the number of hours he played golf on his 5-week holiday.

8 Reah hates TV ads. During a one-hour period she counted there were 12 30-second ads and 24 15-second ads. How many minutes of ads were included in the one-hour period?

9 The Rockets and the Rascals are opposing teams in a rugby league match. Tries are worth 4 points and goals 2 points. The Rockets scored 5 tries and 3 goals and the Rascals scored 6 tries and no goals. Which team won and by how many points?

10 The Boomers and the Snakes are two opposing teams in an AFL match. In AFL, teams score 6 points for a goal and 1 point for a behind. At three-quarter time the Boomers had scored 18 goals and 2 behinds while the Snakes had scored 17 goals and 8 behinds. Who is winning?

11 The BBQ Specialists are catering for Mala's garden party. They are supplying catering and bar service for 42 people. How much will Mala have to pay?

❀**The BBQ specialists**

Catering
Steak and salad with
bread rolls: $18.75 per person

Bar service
Fruit juices and
soft drink: $6.50 per person

GROUP ACTIVITY: THE YEAR 11 PARTY

Budget catering

Price per person
Finger food:.................................$5.60
BBQ steak: $9.20
BBQ sausages:...............................$2.10
BBQ seafood:................................$8.20
Salad: $4.35
Bread rolls: 90c
Deserts:....................................$3.95
Coffee/tea:.................................$1.10
Cheese platters and biscuits: $2.60
Fruit platters:.............................$2.40
Non-alcoholic drinks:.......................$7.40

Toby is responsible for organising the catering for the Year 11 end-of-year party. The organising committee has allocated $4000 to spend on the food and non-alcoholic drinks for the 180 people expected to attend.

Your group's task is to help Toby decide on an interesting menu for food and drink. Remember to stay within your budget.

1.06 Estimating costs

It's difficult to mentally calculate the total cost of the items in your supermarket trolley, but careful shoppers can estimate using approximation techniques.

EXAMPLE 5

Patrick is purchasing the supermarket items shown below: 5 cans of dog food, Caesar salad, blueberries, teabags, strawberries and macadamia nuts. The price of each item is shown.

$2.98
$5.50
$4.59
$6.99
$13.98
$4.98

Dilmah

Scout Kozakiewicz

Patrick has $54 in his wallet. Estimate whether he has enough to pay for his shopping.

Solution

Mentally round each price *up* to the nearest dollar.

Dog food:	$5 \times \$3 = \$15$
Caesar salad:	$6
Blueberries:	$5
Teabags:	$7
Strawberries:	$5
Macadamia nuts:	$14

Estimated price = 15 + 6 + 5 + 7 + 5 + 14
$$= 52$$
The items cost approximately $52.

When we estimated the costs, each estimate was more than the true cost.

The true cost of the items is less than $52.

Patrick has $54.

Patrick has enough money to pay for his shopping.

Write the answer to the question.

Exercise 1.06 Estimating costs

Answer all questions *without* using a calculator.

For Questions **1** to **5**, select the correct answer **A, B, C** or **D**.

1 Which is the best estimate of the value of $9.99 × 8?
    A $72   **B** $80   **C** $90   **D** $100

2 What is the best approximate value of $25.99 + $4.10 + $15.05 + $3.90?
    A $45   **B** $49   **C** $55   **D** $60

3 Gavin ordered 19.5 m of materials at a cost of $29.90 per metre. Estimate the cost of the materials.
    A $60   **B** $90   **C** $600   **D** $900

4 Seth is buying small chocolates that cost 49c each. Approximately how much will 50 cost?
    A $9   **B** $25   **C** $99   **D** $2500

5 Davinia buys bags of nuts and bolts to use in her furniture business. The bags contain between 48 and 51 nuts and bolts. She has 11 bags in her supply cupboard. Approximately how many nuts and bolts does she have?
    A   50   **B**   550   **C**   1100   **D**   2500

6 Christina cuts 18.9 m long lengths of wool for the warp in her loom. She requires 82 lengths.

    a   Approximately how many metres of wool will she need for the 82 warp threads?
    b   The wool Christina uses comes in 400-metre long balls. How many balls of wool will Christina need?
    c   Use a calculator to check your answer to part **b**.

7 When Bryce bought 2 pizzas at $8.95 each and a salad for $5.80, he paid with a $50 note. Approximately how much change should he get?

8 Sebastian has only $60 remaining on his debit card until payday. He has these items in his supermarket basket:

- 2 packets of sausages @ $5.99 per packet
- 1 packet of potatoes $4.99
- 1 carton of milk $3.95
- 1 carton of yoghurt $6.95
- 1 packet of breakfast cereal $5.95

a Estimate the cost of the items in Sebastian's supermarket basket.

b Will he have sufficient funds in his account to pay for his shopping?

c Does he have enough money to add a $9.80 carton of soft drink to his shopping?

9 Paige is knitting a scarf. It takes her 29 seconds to complete 1 row.

a Explain why we can use the calculation $80 \div 2 = 40$ to calculate the number of minutes Paige takes to complete 80 rows of knitting.

b Paige needs to knit 50 more rows to finish the scarf. Approximately how long will it take her?

INVESTIGATION

MY SHOPPING DOCKET

To complete this investigation you will need samples of supermarket shopping dockets.

What you have to do

- Estimate the total cost of the items on a docket.

- Compare your estimate to the total on the docket. Was it more or less than the actual total?

- What is the difference between your estimate and the true cost?

- Use this formula to calculate how much your estimate is different from the true cost as a percentage.

$$\text{Percentage error} = \frac{\text{difference}}{\text{true value}} \times 100\%$$

- Repeat the process for a different shopping docket. Does your percentage error become smaller with practice?

Chapter problem

You've covered the skills required to solve the chapter problem. Can you solve it now?

1.07 Rounding decimals

Calculators often show a lot of **decimal places**. When Imran used his calculator to divide 5.7 by 1.3, it showed the answer as 4.384 615 385. Imran wants the answer with only 1 decimal place, so he has to **round** his answer.

Imran has to decide whether 4.384 615 385 is closer to 4.3 or 4.4. He will check using just the first 2 decimal places: 4.38. On the number line, he can see that 4.38 is closer to 4.4.

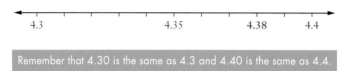

Remember that 4.30 is the same as 4.3 and 4.40 is the same as 4.4.

This is because the 8 in 4.38 is more than 5, which is the halfway mark. Imran will round to 4.4.

When the digit following the decimal place you want to round to is:

- **4 or smaller** – just **round down** and leave all the extra digits off.

- **5 or bigger** – **round up** and add 1 to the digit in the decimal place you want to round to, and leave all the extra digits off.

This table shows the results when some decimals are correctly rounded to 1 decimal place.

Original decimal	Rounded to one decimal place
6.802 956	6.8
6.830 5	6.8
6.849 99	6.8
6.853	6.9
6.893	6.9

EXAMPLE 6

Round 18.3721 to one decimal place.

Solution

Decide whether 18.3721 is closer to 18.3 or 18.4.
The figure following the 1st decimal place is 7.

Round the 3 up to 4. $18.3721 \approx 18.4$.

Exercise 1.07 Rounding decimals

1 Round each number correct to 1 decimal place.

Example 6

   a   16.1256   **b**   29.7681   **c**   14.6472

   d   13.2836   **e**   104.554   **f**   195.219

2 Write 124.727 56 correct to 2 decimal places.

3 Darryn correctly rounded a number to 16.3. Suggest 2 possible values the original number could have been.

4 In surfboard-riding competitions, scores are rounded to 1 decimal place. This is how scores are calculated.

Step 1: Each of 5 judges awards a result from 0.0 to 10.0 for the wave.

Step 2: The lowest and highest scores are removed.

Step 3: The **average** of the 3 remaining scores is determined by adding them together, then dividing by 3. The average is rounded to 1 decimal place.

Step 4: Each competitor's best 2 scores are added to obtain their final score for the competition.

Here are the scores that the judges gave Sally for her first wave. Use calculations to show that Sally's score for her first wave is 7.7.

7.5	7.3	8.1	7.9	7.6

5 Mia and Elissa are competing against each other in a surfboard-riding competition. This table shows the points judges awarded them on 4 waves.

Mia's scores					Elissa's scores			
Wave 1	Wave 2	Wave 3	Wave 4		Wave 1	Wave 2	Wave 3	Wave 4
4.7	7.2	8.5	7.4		6.9	7.8	5.2	8.2
5.1	7.1	8.5	7.8		7.0	7.9	5.1	8.1
4.9	7.5	8.4	7.6		7.2	7.8	5.3	8.4
5.0	7.3	8.2	7.9		6.9	8.0	5.2	8.0
4.9	7.5	8.6	7.5		7.1	7.8	5.0	8.9

   a   Calculate the score awarded to each surfer for each wave.

   b   Calculate each surfer's final score for the competition.

   c   Who won the event?

6 Express the total of 2 m, 148 cm and 384 mm in metres, correct to 1 decimal place.

7 Ben's car travels 8.9 km on 1 L of petrol. The car's fuel tank holds 60.5 L of petrol. How far can the car travel on one full tank of petrol? Express your answer in kilometres, correct to 1 decimal place.

8 Pauline buys knitting wool in **hanks** from a wool mill. Two of the hanks contained 1560 m each, and the other 3 hanks contained 1875 m each.

   a Calculate the total length of wool in Pauline's 5 hanks.

   b How many kilometres of wool are in the 5 hanks? Express your answer correct to 1 decimal place.

9 Thanh bought 79.3 L of petrol at 155.8c/L

   a Explain why the value of 80×1.5 will give the approximate cost of the petrol in dollars.

   b Approximately how much will the petrol cost?

   c Use a calculator to determine the cost of the petrol correct to the nearest 5 cents.

   d How different is your approximation to the actual cost?

10 The gas company replaces the LPG (liquid petroleum gas) in the bottles at Jaye's house each month. Today's delivery docket showed a delivery of 151.65 L of LPG priced at $1.35 per litre.

   a Jaye wanted to estimate the total cost. She calculated that the cost will be between $150 \times \$1$ and $150 \times \$2$. Will the cost be closer to $150 or $300? Give reasons for your answer.

   b Calculate the value of $151.65 \times \$1.35$ and express the answer correct to 2 decimal places.

   c Jaye allows $2200 in her annual budget for gas. Is this too much or not enough? Justify your answer.

11 Simone paid $48.62 for 3 metres of fabric. Calculate the amount she paid per metre, correct to the nearest cent.

ACCURACY IN MEASURING LENGTHS

In this activity you are going to investigate how accurately each member of your group can measure lengths. Each group will need a tape measure that measures in metres, centimetres and millimetres.

What you have to do

1 Choose 3 suitable distances in your school environment; for example, the length of the school verandah, or the distance from your classroom doorway to the nearest tree.

2 Each member of the group measures the lengths as accurately as possible.

3 Compare your group's measurements. At what level of accuracy (for example, answers in metres correct to 1 decimal place) are the measurements the same?

1.08 Practical multiples

Nuts and bolts can be sold in packets of 20 or 100, fertiliser in 40 kg bags and copy paper in lots of 500. Many manufacturers package items in multiple quantities for use by tradespeople, and in most cases it's not possible to buy part of a packet.

EXAMPLE 7

Anna starts a dog-walking service and wants to have some advertising pamphlets printed. The printer handles advertising pamphlets only in multiples of 250. Anna wants 600 pamphlets. How many will she have to order?

Solution

The multiples of 250 mean 'lots of 250', that is 1×250, 2×250, 3×250, 4×250, 5×250, etc. The first 5 multiples are 250, 500, 750, 1000, 1250. Anna can't order 600 pamphlets so she must choose the next multiple of 250 higher.

Anna will have to order 750 pamphlets if she requires 600 pamphlets.

EXAMPLE 8

Timber is available only in lengths that are multiples of 300 mm.

a Is it possible to buy a 1 m length of timber? Why, or why not?

b Peter needs 2 lengths of timber each 1300 mm long. What is the best way for him to buy the timber?

Shutterstock.com/Uber Images

Solution

a Determine whether 1 m is a multiple of 300 mm by dividing it by 300 mm. If the answer is a whole number, then it is a multiple of 300 mm. First convert 1 m to mm.

1 m = 1000 mm

1 m ÷ 300 mm = 1000 mm ÷ 300 mm

$$= 3.333...$$

The answer is not a whole number, so 1 m is not a multiple of 300 mm.

It is not possible to buy a 1 m length of timber.

b Find the total length.

Total length = 2 × 1300 mm

$$= 2600 \text{ mm}$$

Divide by 300 mm to check if it is a multiple of 300.

2600 ÷ 300 = 8.666... (not a multiple of 300).

It isn't so round up to 9 and buy 9 × 300 mm.

Length required = 9 × 300 mm

$$= 2700 \text{ mm}$$

Cut 2 × 1300 mm from 2700 mm length (100 mm left over).

Exercise 1.08 Practical multiples

1 Business envelopes are sold in boxes of 200.

Example 7

  a At the end of each month, Jane posts accounts to all the company's customers. She ordered 8 boxes of envelopes for the letters. How many envelopes are contained in the boxes Jane ordered?

  b Mike requires 600 envelopes to post statements to customers. How many boxes of envelopes will he need?

  c What are the first four multiples of 200?

  d Joel needs 840 envelopes. How many boxes does he need to order?

2 Timber is available only in multiples of 300 mm.

Example 8

  a Explain why you can't buy a piece of timber 1400 mm long.

  b If you need a piece of timber 1400 mm long, what length do you need to buy?

  c Oliver needs 2 pieces of timber each 1150 mm long. What length of timber should he buy?

  d A building plan requires a piece of timber 1.04 m long. What length of timber should be bought?

3 At the wholesaler, electrical wire is available only in multiples of 100 m. Kelly calculated that for her next job she will need 3×74 m lengths of wire and 2×180 m lengths. How much wire does she need to buy for the job?

4 Rose likes to make patchwork quilts. At the shop, fabric is available only in lengths that are multiples of 20 cm. Rose requires 15 cm of a pink fabric and 22 cm of a white fabric. Both fabrics cost $22 per metre.

  a What length of each fabric does Rose need to buy?

  b How much will the fabric that she needs cost?

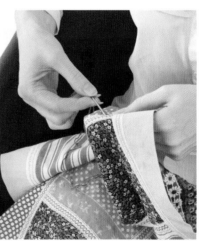

Shutterstock.com/Strakovskaya

5 Darryl is making a fence from treated pine. The wood is available in lengths that are multiples of 0.3 m from 2.1 m to 3 m. He needs 17 posts each 1.35 m long, and 8 top rails each 2.3 m long. What lengths of timber should he order from the timberyard?

1.09 After the point

While 1.5 metres means 1 metre and 50 centimetres, 1.5 hours doesn't mean 1 hour and 5 minutes, and 1.5 years doesn't mean 1 year and 5 months. Fortunately, scientific calculators have a 'degrees, minutes, seconds' key (**o, ,,** or **DMS**) that converts 1.5 hours into 1 h 30 min, but in other situations we have to think for ourselves!

EXAMPLE 9

What does the .45 represent in the following 2 situations?

a $16.45

b   16.45 metres

Solution

a   For money, the 2 digits after the decimal point represent cents.

In $16.45, the .45 represents 45 cents.

b   For metres, the 2 digits after the decimal point represent centimetres.

In 16.45 m, the .45 represents 45 cm.

EXAMPLE 10

Write 2.7 hours in hours and minutes.

Solution

Change 2.7 hours to hours and minutes using your calculator by pressing 2.7 **o, ,,** **=** or 2.7 **2ndF** **DMS**.

2.7 hours = 2 h 42 min

OR first change the decimal part, 0.7 hours, into minutes by multiplying it by 60 minutes (1 hour).

0.7 hours = 0.7×60

$= 42$ minutes

Write the answer.

2.7 hours = 2 h 42 min

NELSON QMATHS 11. Essential Mathematics

ISBN 9780170412650

EXAMPLE 11

Express 1.6 months in months and days, assuming that an 'average month' is 30 days.

Solution

Change 0.6 months into days.	0.6 month = 0.6 × 30
Multiply 0.6 by 30.	= 18 days
Write the answer.	1.6 months ≈ 1 month and 18 days.

Exercise 1.09 After the point

1 What does the .24 represent in each amount?

   a $7.24
                        b   7.24 m

Example 9

2 What does the .9 represent in each measurement?

   a $18.9
                        b   18.9 m

3 Use the ⟨○'"⟩ or ⟨DMS⟩ key on your calculator to express the following times in hours and minutes.

   a   2.5 h   **b**   3.8 h   **c**   1.4 h   **d**   2.9 h

Example 10

4 Follow these steps to convert 4.5 years into years and months.

   a   How many months are there in 1 year?

   b   Multiply the number of months in a year by 0.5.

   c   4.5 years is 4 years and the number of months you calculated in part **b**.

5 Convert 2.25 years into years and months.

6 a   How many months are equivalent to 0.75 years?

   b   Write 5.75 years in years and months.

7 Express each time in months and days. Assume that there are 30 days in a month and write your answers correct to the nearest day.

   a   6.5 months   **b**   8.9 months   **c**   3.24 months   **d**   5.3 months

Example 11

8 A cricket match is measured in 'overs'. An over consists of 6 balls.

   a   Express 18 balls in 'overs'.

   b   During a test match the commentator said 'There are 18.5 overs remaining until the end of play'. If the commentator was using mathematics correctly, how many overs and balls should be remaining?

   c   The commentator meant that there were 18 overs and 5 balls remaining. Why do you think cricket commentators use decimal points in this non-standard way?

CALCULATIONS FIND-A-WORD PUZZLE

Use each clue to find 11 keywords. Then copy the puzzle grid below and find the same 11 keywords in the grid.

1 Multiply by 2.

2 Multiply by 3.

3 The centre of a dartboard

4 A positive or negative whole number

5 The initials we use to remember the 'order of operations' rules

6 Above zero

7 Another word for powers, beginning with I

8 For order of operations, we calculate what's inside these first.

9 Below zero

10 A game played on a circular board numbered with points values.

11 Electronic Funds Transfer at Point Of Sale (initials).

X	P	O	S	I	T	I	V	E	Y	T
D	M	I	C	T	N	P	D	L	S	G
I	R	A	U	B	I	D	M	A	S	W
N	B	N	E	G	A	T	I	V	E	J
T	U	R	W	Z	C	F	R	C	J	A
E	L	O	A	K	T	S	X	N	E	I
G	L	D	M	C	L	R	H	C	Q	S
E	S	H	Y	W	K	G	I	V	O	B
R	E	S	A	J	R	E	F	P	O	Y
B	Y	E	H	D	A	R	T	S	L	R
K	E	L	P	O	X	F	N	S	R	E
D	O	U	B	L	E	M	F	O	Z	U

SOLUTION TO THE CHAPTER PROBLEM

Problem

Ben bought fish and chips for $11.25 and paid with a $50 note. The cashier gave him a $20 and a $5 note and a few coins for his change. Could the change be wrong?

Solution

The fish and chips cost approximately $10. Ben's change should have been approximately $50 − $10 = $40. Ben's change included a $20 and a $5 note plus a few coins, so the change was probably between $25 and $30.

Yes, he has probably been short-changed (by about $10).

1. CHAPTER REVIEW

What's the score?

How well have you mastered the skills in this chapter? Test yourself with these questions.

1 Add each set of dartboard scores without using a calculator.
  a A 5 and a 4
  b Double 5 and a 3
  c Double 3 and double 8
  d A 5 and a triple 6

Exercise 1.01

2 Jasper is playing darts. Calculate his score when his 3 darts landed on:
  a 5, 7 and double 11
  b double 9, 6 and triple 20

Exercise 1.02

3 Mike and Tony are playing golf. Mike's final score is −3 and Tony's is +2. Who won the game and by how many strokes?

Exercise 1.03

4 Evaluate each expression without using a calculator, then use a calculator to check your answer.
  a $8 + 5 \times 2$
  b $20 \div 10 + 3 \times 4$
  c $(15 + 9) \div (8 - 2)$

Exercise 1.04

5 Use a calculator to evaluate each expression correct to 1 decimal place.
  a $17.35 + 182.96 - 25.47$
  b $16.39 + 12.5 \div 3.6$
  c $(12.3)^2$

Exercise 1.04

6 Jemima bought 8 balls of wool priced at $11.99.
  a Estimate how much the wool costs.
  b The shop charged her $71.94. Is the bill right?

Exercise 1.05

7 Isabella ordered the food for her dance group. She ordered 4 chicken wraps and 6 small sushi. The chicken wraps cost $5.50 each and the sushi cost $2.90 each. She paid with a $50 note. How much change should Isabella get?

Exercise 1.06

8 **a** Round 15.32 up to the nearest whole number.
  b Round 16.75 down to the nearest integer.

Exercise 1.07

9 Round each measurement correct to the nearest whole number.
  a 27.9 kg
  b 135.2 m
  c 4.5 mm

Exercise 1.07

10 The buttons that Mai wants to use on a jacket she's designing are sold in multiples of 3. Mai needs 13 buttons for the jacket. How many buttons will she have to buy?

Exercise 1.08

11 Approximately how many days are there in 4.7 months?

Exercise 1.09

2.

GIVING 110%

Chapter problem

Tanika sells cosmetics. She earns commission at the following rates.

Commission on Tanika's monthly sales	
First $500 of sales	5%
On the next $1000	4%
Remainder of sales	3.5%

Calculate Tanika's commission in a month when her sales were $4200.

WHAT WILL WE DO IN THIS CHAPTER?

- Calculate a percentage of an amount
- Express one amount as a percentage of another
- Increase and decrease an amount by a percentage
- Solve problems involving cost price, selling price, profit, loss, discounts and GST

HOW ARE WE EVER GOING TO USE THIS?

- Calculating discounts on items we want to buy
- Calculating statistics in sport
- Whenever we come across a percentage – which is often!

2.01 Finding a percentage of a quantity

Fractions, decimals and percentages

$$\text{Percentage of a quantity} = \frac{\text{Percentage}}{100} \times \text{quantity} \quad \text{OR}$$
$$= \text{Percentage} \div 100 \times \text{quantity}$$

Percentage shortcuts

EXAMPLE 1

Calculate:

Percentages without calculators

a   23% of $1650

b   75% of 620 kg    'of' means multiply.

Solution

Percentage calculations

a   Write the percentage as a fraction or decimal and then multiply by the quantity.

$23\% \text{ of } \$1650 = \dfrac{23}{100} \times \$1650$

or $0.23 \times \$1650$

$= \$379.50$

b   Write the percentage as a fraction or decimal, then multiply by the quantity.

Some calculators have a percentage key. Check with your teacher whether it is practical to use it.

$75\% \text{ of } 620 \text{ kg} = \dfrac{75}{100} \times 620$

$= 465 \text{ kg}$

Mental calculations

Simplifying fractions

EXAMPLE 2

Find 32% of 4 m. Give your answer in cm.

Sometimes it is necessary to change the units of the quantity before we calculate the percentage.

Solution

Changing units

Change 4 metres to centimetres.

$4 \text{ m} = 4 \times 100 \text{ cm}$

$= 400 \text{ cm}$

Write the percentage as a fraction or decimal and then multiply by the quantity.

$32\% \text{ of } 400 = 0.32 \times 400$

$= 128 \text{ cm}$

Exercise 2.01 Finding a percentage of a quantity

1 Find:

a	11% of $300	**b**	60% of 140 kg	**c**	42% of 600 cm
d	25% of $1230	**e**	90% of 240 marks	**f**	85% of 700 000 people
g	19.4% of 785 kg	**h**	2.5% of 200 m	**i**	35% of 840 students
J	5.9% of $5300	**k**	12.5% of 96 L	**l**	17.75% of 640 hectares

2 Calculate each amount.

a   12% of 8 m, giving your answer in cm

b   35% of 10 days, giving your answer in hours

c   60% of 1 year, giving your answer in days

d   15% of 26 L, giving your answer in mL

e   10% of $2.30, giving your answer in cents

f   87.5% of 16 weeks, giving your answer in days

g   22.5% of 6 km, giving your answer in metres

h   40% of 1.5 tonnes, giving your answer in kg

3 20% of the students at Nelson State High School study Chinese. There are 1290 students at the school. How many study Chinese?

4 Antony earns 5.5% interest on an investment account. He has $7500 in the account. How much interest will Antony earn in one year?

5 Employees at the GMC Steel Company are given a 2.5% pay increase.

a   Dinesh is paid $32.40 per hour. How much extra will he be paid per hour after the increase?

b   Sofija is paid $82 500 per year. How much extra will she be paid per year after the increase?

6 Xi's laptop can be used for 6 hours without recharging. Her laptop shows she has 62% of her time left. How much time does she have left before her laptop needs recharging?

7 Michael planted 850 seedlings in his market garden. 92% of the seedlings survive to full maturity. How many seedlings survive?

8 When Frankie orders stationery for her business, she must pay 10% GST. Calculate the GST payable on stationary purchases of $524.

9 In a poll of 1500 people, 85% of people agreed 'that Mathematics is the most useful subject to study at school'. How many people is this?

10 At the Strolling Bones concert, 65% of seats need to be sold for the promoter to make a profit. The venue seats 18 000 people. How many seats need to be sold to make a profit?

11 Jason decides to go on a weight loss program. His starting weight is 187 kg. In the first six weeks he loses 13% of his body weight. How many kilograms did Jason lose in the first six weeks?

12 Taylor is downloading the latest album of her favourite band. The file is 750 megabytes. So far she has downloaded 45% of the album. How many megabytes has Taylor downloaded so far?

Chapter problem

You've covered the skills required to solve the chapter problem. Can you solve it now?

2.02 What percentage of ...?

We often change amounts into percentages so that we can compare them. For example, we calculate our marks in tests as percentages so we can compare our performance in each subject. To do this, we must first write one amount as a fraction of another, then we change the fraction to a percentage by multiplying it by 100%.

$$\text{Percentage} = \frac{\text{amount}}{\text{whole amount}} \times 100\% \qquad \text{OR}$$

$$= \text{amount} \div \text{whole amount} \times 100\%$$

EXAMPLE 3

18 out of the 25 students in the class have part-time jobs. What percentage of the class is this?

Solution

Write as a fraction and multiply by 100%. $\frac{18}{25} \times 100\% = 72\%$

Write your answer.   72% of the class have part-time jobs.

When we compare two amounts by writing one as a percentage of the other, we need to make sure that they are in the same units.

EXAMPLE 4

Express 36 minutes as a percentage of 2 hours.

Solution

Both quantities should be in the same units, so change 2 hours to minutes.	2 hours = 2 × 60 minutes
	= 120 minutes
Write as a fraction and multiply by 100%.	$\dfrac{36}{120} \times 100\% = 30\%$
Write your answer.	36 minutes is 30% of 2 hours.

Exercise 2.02 What percentage of ... ?

1 Convert each test mark to a percentage.

   a   32 out of 40   **b**   11out of 20   **c**   60 out of 75   **d**   71 out of 80

Example 3

2 In a driving test, Melanie answered 21 out of 24 questions correctly.

   a   What percentage is this?

   b   The pass mark on the driving test is 95%. Did Melanie pass the test?

3 Joe earns $1005 per week and pays $420 in tax. What percentage of his earnings does Joe pay in tax? Answer correct to one decimal place.

4 Express each measurement as a percentage. Remember to have both amounts in the same units! Answer correct to one decimal place.

Example 4

a	5 minutes of 1 hour	**b**	18 hours of 4 days	**c**	75c of $6
d	75 mm of 20 cm	**e**	520 kg of 1 tonne	**f**	400 mL of 3.5 L
g	3 months of 1 year	**h**	3 days of 1 fortnight	**i**	50 days of 1 year
j	25 minutes of 3 hours	**k**	36 hours of 1 week	**l**	5 fortnights of 3 years

5 In a survey of 1500 people, 1125 of the respondents agreed that breakfast is the most important meal of the day. What percentage is this?

6 Brianne sells jeans in her boutique for $85. She increases the price by $15. What percentage increase is this? Answer correct to one decimal place.

7 Australia has an area of 7.7 million square kilometres. Of this, desert takes up 3.8 million square kilometres. What percentage of Australia is desert? Answer correct to one decimal place.

8 The cost of Ziad's new glasses is $295. His health fund pays him a benefit of $165. What percentage is Ziad's benefit of the cost of his glasses? Answer correct to one decimal place.

9 Vanessa buys 6 metres of rope to tie up her plants. She uses 315 cm in her backyard. What percentage of the rope has she used?

10 George and Sue are travelling to see their parents. The trip takes 7 hours. They take a break after 2 hours 40 minutes. What percentage of the trip have they completed when they take the break? Answer correct to one decimal place.

11 In a senior class of 23 students, 8 walk to school and 11 travel by bus. Calculate, correct to one decimal place, the percentage of students who:

a walk to school   **b** travel by bus

12 Kyle had the following results in his half-yearly exams.

English: 45 out of 75

Mathematics: 38 out of 70

Science: 80 out of 125

a Calculate Kyle's percentage in each subject. Answer correct to one decimal place.

b In which exam did Kyle achieve the highest result?

HOW MUCH SUGAR IS IN THE FOOD YOU EAT?

Sugar is one of the contributing factors to weight gain. In this investigation you will examine the percentage of sugar in what you eat and drink.

Did you know that sugar has many different names? Some of them include:

- Sucrose – normal table sugar

- Corn sugar – made from cornstarch

- Fructose – sugar in honey, plants and some fruit

- Lactose – found in milk

- Mannitol – a sugar alcohol in many plants

What you need to do

1 Choose 10 items that you regularly eat or drink. You will need to choose items that have product information on the packaging. You should include a soft drink, cereal, pasta or rice, some canned food and a variety of other foods.

2 For each item, find:

   a the serving size

   b the amount of sugar per serving

   c the percentage of sugar in the product

3 Which product had the highest percentage of sugar? Which had the lowest? Write a list of the items from most sugar to least sugar.

4 Compare your findings with other students in your class.

5 Write a report on sugar levels in food. Include recommendations about how a person could reduce the amount of sugar in their diet.

Nutrition Facts

Serving Size 2/3 cup (51g)
Servings Per Container About 9

Amount Per Serving	Cereal	Cereal with 1/2 cup Skim Milk
Calories	240	280
Calories from Fat	70	70

	% Daily Value**	
Total Fat 8g*	**12%**	**12%**
Saturated Fat 2.5g	**13%**	**13%**
Trans Fat 0g		
Cholesterol 0mg	**0%**	**0%**
Sodium 50mg	**2%**	**5%**
Total Carbohydrate 37g	**12%**	**14%**
Dietary Fiber 3g	**12%**	**12%**
Sugars 13g		
Protein 4g	**8%**	**16%**
Vitamin A	0%	4%
Vitamin C	0%	0%
Calcium	2%	15%
Iron	6%	6%

iStock.com/CHRISsadowski

2.03 Percentage increase

Percentage increase means to make a quantity bigger by a given percentage, for example, in pay increases or adding GST to prices.

EXAMPLE 5

Increase $500 by 12%.

There are two ways to do this. Choose the method you like!

Solution

Find 12% of $500.	$0.12 \times 500 = \$60$
Add this amount to $500.	$\$500 + \$60 = \$560$
OR New amount is 100% plus 12%.	$100\% + 12\% = 112\%$
Find 112% of $500.	$1.12 \times 500 = \$560$

EXAMPLE 6

All Government employees receive a 2.5% pay increase. Georgie is on a salary of $82 000 per year. What is her new salary?

Solution

Find 2.5% of $82 000.	$0.025 \times 82\,000 = \$2050$
Add this amount to $82 000.	$\$82\,000 + \$2050 = \$84\,050$
OR New amount is 100% plus 2.5%.	$100\% + 2.5\% = 102.5\%$
Find 102.5% of $82 000.	$1.025 \times 82\,000 = \$84\,050$
Answer the question.	Georgie's new salary is $84 050.

Exercise 2.03 Percentage increase

1 a Increase 95 kg by 60%.   **b** Increase $2500 by 6%.

   c Increase 150 m by 5%.   **d** Increase 10 L by 33%.

2 Increase a train fare of $7.50 by 20%.

3 Increase a price of $620 by 10%.

4 Nathan earns $32.75 per hour. His boss gives him a pay increase of 8%. What is his new hourly rate?

5 Saria runs a fashion business. She determines the selling price of each item of clothing by increasing the cost price by 95%. Find the selling price of a jacket that costs Saria $120.

6 Non-essential items in Australia attract a GST (goods and services tax) of 10%. Find the price of a car costing $18 900 after GST is added.

7 Ken and Amanda buy a house for $64 000 and sell it 10 years later at a profit of 147%.

   a Calculate the profit.

   b What is the selling price of the house?

8 Binnsfield Council plans to increase the amount of parkland in the area by 20%. At present there are 20 hectares of parkland. How much parkland will there be under the new plan?

9 Marcus' weekly pay is $945. He is paid an 8% bonus for 2 weeks. How much will he be paid for these 2 weeks?

10 La Plage restaurant charges a 15% surcharge on public holidays. Find the total charge for a meal costing $160 on Australia Day.

11 The Luxury Hotel charges an additional 1.5% when the customer pays with a credit card. Find the total cost of 3 nights' accommodation at $185 per night if it is charged to a credit card. Round your answer to the nearest cent.

12 Billie practises the trumpet for 4 hours each week. In the weeks before competitions, she increases her practice time by 30%.

   a How long does Billie practise the trumpet per week in the weeks before competition?

   b Calculate the average time Billy practises her trumpet per day in the weeks before competition.

13 Jayden buys TVs at a cost price of $295. He adds 75% of the cost price to get his selling price, then he adds 10% GST to calculate the final selling price.

   a Find the price of the TV after Jayden has added his 75% markup.

   b Find the price of the TV after he adds GST. Round your answer to the nearest cent.

   c Is it likely that Jason will charge his customers exactly that price? Why?

2.04 Percentage decrease

Percentage decrease means to make a quantity smaller by a given percentage, for example, in discounts and sale prices.

EXAMPLE 7

Decrease $340 by 7%.

Solution

Find 7% of $340.	$0.07 \times 340 = \$23.80$
Subtract this amount from $340.	$\$340 - \$23.80 = \$316.20$
OR New amount is 100% minus 7%.	$100\% - 7\% = 93\%$
Find 93% of $340.	$0.93 \times 340 = \$316.20$

EXAMPLE 8

During an end-of-financial-year sale, computer games are discounted by 25%. Find the cost of a game regularly priced at $115.

Alamy Stock Photo/René van den Berg

Solution

Find 25% of $115.	$0.25 \times 115 = \$28.75$
Subtract this amount from $115.	$\$115 - \$28.75 = \$86.25$
OR New amount is 100% minus 25%.	$100\% - 25\% = 75\%$
Find 75% of $115.	$0.75 \times 115 = \$86.25$
Answer the question.	The computer game will cost $86.25.

Exercise 2.04 Percentage decrease

1 **a** Decrease $150 by 57%   **b** Decrease 2000 L by 28%

c Decrease 110 kg by 4.5%   **d** Decrease 840 students by 15%

e Decrease 8 hours by 10%   **f** Decrease 4 weeks by 25%

2 Decrease a price of $330 by 25%.

3 Decrease a town's population of 32 780 by 5%.

4 In the January sales all whitegoods are discounted by 15%. Jenna buys a new washing machine usually priced at $799. Calculate the amount Jenna pays for the washing machine during the sale.

iStock.com/billyfoto

5 Last year the Australian road toll was 1603 deaths. This year the road toll decreased by 10.3%. Calculate the road toll for this year. Round to the nearest whole number.

6 Keira and her sister, Amber, have singing lessons. Keira pays full fees of $450 per term, but Amber gets a 5% family discount.

a Calculate how much Amber is charged for her lessons.

b Calculate the total amount of fees paid for the 2 sisters.

7 The student population at Binnsfield Christian College in 2016 was 879. This decreased by 7.5% the following year. How many students were at Binnsfield Christian College in 2017? Round to the nearest whole number.

8 An electricity company offers a 2% discount when customers pay the account on time.

    a Calculate the discount if a bill of $1978.70 is paid on time. Round your answer to the nearest cent.

    b How much does the customer pay?

9 Katie works at the local supermarket. She is given a staff discount of 3.5% on all purchases. Find how much she pays for purchases totalling $178.25.

10 At the 'End of model' sale, Honest John's car yard offers a discount of 30% on all cars. Find the cost of a family car originally priced at $39 990.

11 For tax purposes the value of a computer is depreciated at a rate of 20%. Samir buys a computer priced at $1290. What will be its depreciated value in his next tax return?

12 Thomson's Hardware is having a Christmas sale. All items in the store are discounted by 18%.

    a Lauren is a painter. She buys ten 4 L tins of paint regularly priced at $67 per tin.

        i Calculate the cost of the paint at its regular price.

        ii Calculate the cost of the paint after the 18% discount.

    b Because she is a painter, Lauren is also entitled to a trade discount of 10% of the discounted price. Calculate the cost of the paint for Lauren.

2.05 Profit and loss

All successful businesses make a **profit**. Businesses that make a **loss** quickly close down. Most businesses express the profit or loss they make as a percentage of their cost price.

$$\text{Percentage profit} = \frac{\text{profit}}{\text{cost price}} \times 100\%$$

$$\text{Percentage loss} = \frac{\text{loss}}{\text{cost price}} \times 100\%$$

The **cost price** is how much a business buys the item for, from the supplier.

The **selling price** is how much a business sells the item for, to the customer.

EXAMPLE 9

Jacinta buys a necklace for her market stall for $36. She sells it for $47.

a How much profit does Jacinta make when she sells the necklace?

b Calculate the percentage profit. Answer correct to 1 decimal place.

Shutterstock.com/Grisha Bruev

Solution

a Profit = selling price − cost price.

Profit = $47 − $36

= $11

b Write the profit over the cost price and change it to a percentage by multiplying by 100%.

$$\text{Percentage profit} = \frac{11}{36} \times 100\ \%$$

$$= 30.5555\ ....\%$$

$$\approx 30.6\%$$

EXAMPLE 10

During a sale, a bed is sold for $599. The bed cost the store $750.

a Calculate the loss for the store on the bed.

b Calculate the percentage loss, correct to 1 decimal place.

Alamy Stock Photo/Robert Stainforth

Solution

a Loss = cost price − selling price.

$$\text{Loss} = \$750 - \$599$$
$$= \$151$$

b Write the loss over the cost price and change it to a percentage by multiplying by 100%.

$$\text{Percentage loss} = \frac{151}{750} \times 100 \%$$
$$= 20.13333 \ldots \%$$
$$\approx 20.1\%$$

Exercise 2.05 Profit and loss

1 Calculate, correct to one decimal place, the percentage profit for each sale.
  a cost price $35, selling price $45
  b cost price $625, selling price $810
  c cost price $149.75, selling price $274.25
  d cost price $2.60, selling price $4.05

2 Calculate, correct to one decimal place, the percentage loss for each sale.
  a cost price $516, selling price $420
  b cost price $2175, selling price $1300
  c cost price $12 500, selling price $11 000
  d cost price $5.20, selling price $4.40

3 Carrie pays $40 for a small cupboard at a garage sale. She later sells it for $75 to a furniture dealer. Calculate her percentage profit.

4 Paolo pays $165 for a ticket to the football grand final. When his team doesn't make the grand final, he sells the ticket for $90. Calculate his percentage loss.

5 Natasha buys an old bicycle for $20. She spends $115 replacing some parts and painting it. She sells it for $150.

   a Has she made a profit or a loss? How much is the profit or loss?

   b Calculate the percentage profit or loss.

6 Keiran buys a used car for $2500 and spends $1750 replacing some parts. At the end of the year he sells it for $3750.

   a Does he make a profit or a loss? How much is the profit or loss?

   b Calculate the percentage profit/loss.

   c Comment on whether you think Keiran would be happy with this outcome.

7 Matt needs to make 65% profit on the items in his camping store to cover costs. What price does he charge for hiking boots that cost him $165?

8 Sophie buys 2000 shares valued at $23 400. She pays a stockbroker's fee of 2.5% on the value of the shares. She later sells all the shares for $10.55 per share.

   a Calculate the stockbroker's fee for Sophie's purchase.

   b Calculate the total amount she paid for the shares.

   c How much money does Sophie receive for her shares when she sells them?

   d Has she made a profit or a loss on the shares?

   e Calculate the percentage profit or loss on the shares. Answer correct to the nearest whole number.

9 Gianni buys 7 boxes of bananas for $45 each. Each box contains 10 kg of bananas. He sells 4 of the boxes at $5.10 per kg. He reduces the price to $4 per kg to clear the remaining boxes before the fruit goes bad.

   a How much do the bananas cost Gianni?

   b How much does he receive from selling the bananas?

   c Does Gianni make a profit or a loss? Calculate the percentage profit or loss, correct to one decimal place.

10 A shoe shop bought pairs of shoes for $48 each and priced them at $90 per pair. During the end-of-season sale, the manager reduced the price by $20. Calculate the percentage profit the shop makes when it sells the shoes during the sale. Answer correct to 1 decimal place.

Percentage
problems

Percentage
power

Working with
percentages

2.06 Practising percentages

In this exercise you will use all the skills you have learned in this chapter to solve problems involving percentages.

Exercise 2.06 Practising percentages

1 Nelson College's hockey team won 12 games out of 15. What percentage is this?

2 55% of the students at one university are female. The university population is 85 300. How many *males* attend the university?

3 A muesli bar is 35 g.

   a It contains 6.9 g of fat. What percentage is this? Answer to one decimal place.

   b The muesli bar is 64% carbohydrate. How many grams is this?

4 Luka, a real estate agent, earns 2.5% commission on a sale of a house for $345 000. How much was his commission?

5 Zoe buys a used car for $14 000 and sells it at a loss of 20%. Find its selling price.

6 Lachlan's pay rate of $18.45/hour is increased to $19.85/hour. Find the increase as a percentage of the original pay rate. Answer correct to one decimal place.

7 Sarah earns $2346 per fortnight. Her rent is 30% of her income. How much rent does Sarah pay?

8 Alfredo's Spaghetti Sauce contains 11.9% sugar, according to the label. How many grams of sugar are in a 580 g jar?

9 Ashleigh sells kitchenware by visiting homes and is paid a retainer of $320 per week to cover her expenses, plus a commission of 15% on all sales. Ashleigh's sales last week totalled $2896. Calculate her pay for that week.

10 Jacob is buying a house that was priced at $345 000 last year. House prices have risen by 1.5% since then. How much does the house cost now?

11 When a new model is coming, car dealers offer discounts on the current model. Tony is interested in buying a car which has a price of $36 990. The car dealer offers him this car for $31 990. What percentage discount is this?

12 Kaitlyn runs Computer R Us. She purchases a batch of laptop computers that cost her $395 each. She calculates their selling price by adding on 75% of the cost price. What is the selling price of each laptop computer?

13 Daniel's watch loses 3 minutes every 6 hours.

   a What percentage time loss is this?

   b He sets his watch correctly at 8 a.m. in the morning. He finishes work at 5 p.m. in the afternoon. What time will his watch show when he finishes work?

14 So far this season the Cowboys team has scored 584 points and had 355 points scored against them. Calculate their points percentage using the formula below. Answer correct to 1 decimal place.

$$\text{Points percentage} = \frac{\text{points for}}{\text{points against}} \times 100$$

15 The Costello family donates 2% of their annual income to medical research into dementia. Calculate how much they donated in a year when the family income was $225 000.

16 A 500 g can of chicken and corn soup contains 110 g of corn. What percentage of the soup is corn?

17 Shanice buys her wedding dress for $1850. Several months after her wedding, she sells the dress on the Internet for $1280. Calculate her percentage loss on the wedding dress.

18 The book Anthea wrote sells for $54 a copy. She earns a 10% royalty on the first 4000 copies sold and a 12.5% royalty on the remaining sales. Calculate the royalty Anthea receives on the sales of 7000 copies of her book.

INVESTIGATION

USES OF PERCENTAGES

1 Look on the Internet, newspapers, magazines and brochures for examples of different ways we use percentages in real life. Find photographs or draw situations where percentages are used. Present the information you have found as a 1-page display or as a PowerPoint presentation.

2 Select an example from what you have found in part 1. Write a set of 10 questions for the example, using what you have learned about where and how we use percentages in real life. Calculate the answers.

3 Swap the questions you have written with others in the class. Each person should answer at least 3 sets of questions. Ask the student who prepared the questions to mark your answers. If there are any disagreements, check with your teacher.

KEYWORD ACTIVITY

Here are some keywords used in this chapter.

commission	cost price	decrease	discount
GST	increase	interest	loss
profit	selling price		

1 Some of these words are opposites. Find 2 pairs of words that are opposites. Write a short paragraph that shows the meaning of each word.

2 Which word is a way of earning money? Write a sentence explaining its meaning.

3 Describe the difference between *cost price* and *selling price*.

4 Write a sentence explaining the meaning of discount, GST and interest.

SOLUTION TO THE CHAPTER PROBLEM

Problem

Tanika sells cosmetics. She earns commission at the following rates.

Commission on Tanika's monthly sales	
First $500 of sales	5%
On the next $1000	4%
Remainder of sales	3.5%

Calculate Tanika's commission in a month when her sales were $4200.

Solution

Commission on first $500 = 5% of $500

$$= \$25$$

Commission on next $1000 = 4% of $1000

$$= \$40$$

Remainder of sales = $4200 − $500 − $1000

$$= \$2700$$

Commission on remainder = 3.5% of $2700

$$= \$94.50$$

Total commission = $25 + $40 + $94.50

$$= \$159.50$$

Giving 110%

1 Find:

   a 8% of $400

   b 75% of 280 mL

   c 12.5% of 90 hectares

   d 20% of 3 m, giving your answer in cm

   e 72% of 5 kg, giving your answer in g

   f $7\frac{1}{2}$% of 18 h, giving your answer in mins

2 In one year, 42% of all road deaths occurred on country roads. If the road toll for that year was 650, how many people died on country roads that year?

3 Maroun earns 3.5% p.a. interest on her investment of $27 000. How much interest does she earn in one year?

4 Ian scores 80% on his driving theory test consisting of 20 questions. How many questions did he get right?

5 Express as a percentage.

   a Test mark of 34 out of 40

   b Tax payment of $310 out of $1240

   c 3 students out of 30

   d 750 mL of 1 L

   e $17.50 of $70

   f 1500 m of 10 km

6 Louise sells her home for $825 000. The real estate agent is paid a commission of $28 875. What percentage of the sales price is this?

7 Nelson Stadium holds 40 000 people. Recently, 37 200 people attended a finals game at the stadium. What percentage attendance was this? Give your answer correct to 1 decimal place.

8 Angelo has 3 hours to complete an exam. He checks the time after completing 3 questions and he has used 1 hour and 5 minutes of the exam time. What percentage is this? Give your answer correct to 1 decimal place.

9 Increase:

   a $350 by 5%   **b** 60 kg by 55%   **c** 15 000 people by 20%

10 Michael receives a pay increase of 2.5%. Currently he earns $1442 per week. Calculate his new weekly wage.

11 Goods and services tax in Australia is 10%. Find the GST-inclusive price of a television advertised as '$2100 excluding GST'.

12 Decrease:

   a $90 by 40%   **b**   75 m by 15%   **c**   8 weeks by 25%

13 Max buys a shirt at a '40% off all weekend' sale. If the shirt is marked at $75, how much does Max pay for it?

14 The population of Nelsonville was 74 200 people. It has dropped by 7%. Find the current population.

15 Calculate the percentage profit or loss for each situation, correct to 1 decimal place where necessary.

   a   cost price $85   selling price $102

   b   cost price $2500   selling price $3500

   c   cost price $19   selling price $14

   d   cost price $1400   selling price $1267

16 Joanna buys souvenirs for her store at $17 each. She sells them for $29 each. Calculate her percentage profit, giving your answer correct to 1 decimal place.

17 Emad buys shirts for $35 each and sells them for $105 each. In the January sales he reduced the price by $30. Calculate his percentage profit on the sale price.

18 64% of workers at a fast food restaurant are males. How many female workers are there if 325 people work at the restaurant each week?

19 A muesli bar has a serving size of 40 g.

   i   It has 12.2 g of fat. What percentage is this? Answer correct to 1 decimal place.

   ii   It is 5% protein. How many grams of protein are in the muesli bar?

20 Currently, the GST is 10%. Some politicians think it should be raised to 15%. A tradesman charges $320 excluding GST for a morning's work. Calculate how much more you would pay if the GST was raised to 15%.

3.

SOLVING PROBLEMS

Chapter problem

Caroline, Sandra and Sue are entering a craft competition and they have decided to share the costs equally. Sandra paid $102.35 for the material, Sue paid $67.20 for the equipment and Caroline paid $20 for the postage. Who should pay whom and how much should they pay?

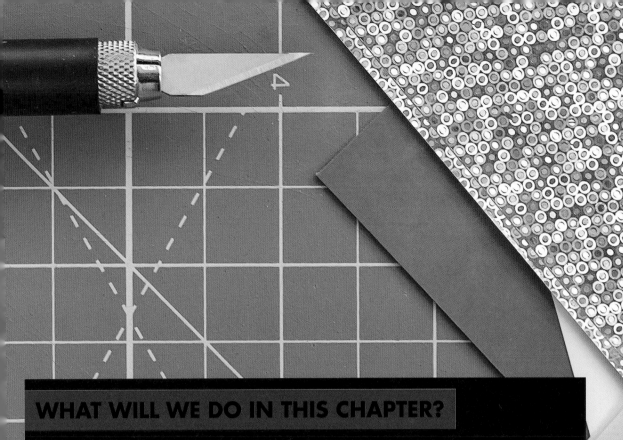

WHAT WILL WE DO IN THIS CHAPTER?

- Interpret a problem and identify what it is
- Identify word clues in the problem that suggest which mathematical operation to use
- Solve the problem using a number of strategies
- Check the solution to the problem to see whether it looks right
- Present the solution to the problem

HOW ARE WE EVER GOING TO USE THIS?

- Can I afford to buy this?
- How much will it cost?
- What time will I have to get there?
- Is the change right?
- How much of each ingredient do I need?

3.01 Does the answer look right?

To be good problem solvers, we need to develop some special skills. We need to be able to:

- identify information in a problem
- change words into mathematical symbols
- identify and use appropriate ways to solve the problem
- spot a wrong or silly answer
- write our answer in a way that others can understand

In this chapter we will learn problem-solving skills, one at a time, starting with spotting a wrong or silly answer.

EXAMPLE 1

The answer to the following problem is not right. What's wrong with the answer?

Problem: Gerry earns $720 per week. How much does he earn per fortnight?

Answer: Gerry earns $650 per fortnight.

Solution

$650 is too small. Fortnights are longer than weeks.

The answer is wrong because it should be should be *more* than $720, not less.

EXAMPLE 2

When Alison bought a cup of coffee, she paid with a $20 note and she received $14.80 change. Could the change be right?

Solution

$20 – $14.80 is about $20 – $15 = $5.

A cup of coffee could cost about $5.

The change could be right.

$5 could be the price for a cup of coffee.

Exercise 3.01 Does the answer look right?

Each answer given in Questions **1** to **8** below is wrong. Explain how you can tell that the answer is wrong without actually calculating it.

1 Sam earns $14 per hour working in a fast food shop. How much will she earn for a 3-hour shift? Answer: $17

Example 1

2 Jake's debit card is linked to his bank account that currently has a balance of $127. He bought a T-shirt for $15 with his debit card. What is the balance of Jake's debit card after he bought the T-shirt? Answer: $142

3 Lara catches the bus to work. The bus trip takes 65 minutes and she starts work at 9 a.m. What is the latest time that she can catch the bus and be at work on time? Answer: 8:15 a.m.

iStock.com/PeopleImages

4 Stuart spends 45 minutes at the gym 5 days per week. How many hours does he spend at the gym each week? Answer: 6 hours

5 A 500g bag of macadamia nuts costs $17.50. At this rate how much does 1 kg of macadamia nuts cost? Answer: $65

6 Lina's car uses 10 L of petrol to travel 100 km. How many litres of petrol will she use to drive 70 km? Answer: 7 km

7 Dimitri has $5000 in a bank savings account that pays 1% p.a. interest. How much interest will he earn for 3 years? Answer: $150 000

8 Convert 60 cm into mm. Answer: 6 mm

Each answer given in Questions **9** to **14** below could be wrong or could be right. Explain how you can tell whether they could be wrong or right.

9 A pay TV company charges customers an additional 1% if they pay their bill by credit card. Sally's bill is $135. How much will she have to pay if she pays her bill by credit card? Answer: $133

10 An electricity supply company gives customers who pay their bill early a 6% discount. Willy's electricity bill is $160. How much will he pay if he pays the bill early? Answer: $150

11 In a survey of 800 people, 550 said they drank tea or coffee with their breakfast. What percentage is this? Answer: 46%

12 Electricity prices are going up by 10%. At the moment, off-peak electricity costs 24c/kWh. How much will off-peak electricity cost after the increase? Answer: 26.4c/kWh

13 It costs Jane $10 to make a bracelet and she sold it for $15. Calculate her percentage profit. Answer: 150%

Shutterstock.com/Ekaterina Shakhova

14 Anthea is going to buy an apartment. In the last year, apartment prices have increased by 10%. Today the apartment costs $220 000. How much did it cost last year? Answer: $200 000

3.02 What do we know?

When we're solving a problem, we need to identify what it is we have to work out and what information we have to do it. Sometimes, we don't have enough information and we have to find more information or make assumptions. Sometimes, we have *too much* information.

EXAMPLE 3

For the problem below, identify:

- what we have to work out
- what information is given to work out the answer.

Kirsty's problem

When Kirsty weighed herself, the scales showed 65 kg. She went on an unhealthy crash diet and lost 12% of her body weight. Calculate Kirsty's weight after her diet.

Solution

The key information is bolded below:

When Kirsty weighed herself, the scales showed **65 kg**. She went on an unhealthy crash diet and **lost 12%** of her body weight. Calculate **Kirsty's weight after her diet**.

Work out: weight after diet.

The information given is:

- Starting weight: 65kg.
- Lost 12%

Exercise 3.02 What do we know?

Keep a record of your answers in this exercise. You will need them for Exercises 3.04 and 3.05.

1 Think about the following problem:

Danica was on holidays in Fiji for 12 days, but it rained on 25% of the days. How many fine days did Danica have in Fiji?

a What is the problem asking us to work out?

b What information is in the problem that can help us solve it?

Example 3

In Questions **2** to **6**, identify what the problem is asking us to work out and what information we know. You do **not** have to solve the problem.

2 Mark wants to buy the 'perfect car'. The recommended price was $36 000, but when he went to buy it, Mark discovered that the dealer had increased all the prices in the caryard by 5%. What is the new price of Mark's 'perfect car'?

3 Charlotte is going to the doctors and she wants to check that she has enough money in her EFTPOS account. How much will she have to pay? The consultation costs $72, but her doctor gives a 5% discount to patients who pay with EFTPOS.

4 Jonah increased the price of all items in his shop. He increased the price of jeans to $54 from $50. By what percentage did Jonah increase prices?

5 Grandma had heart surgery on 12 March. She has to go back to the doctor for a check-up in 8 days and then 6 weeks after the surgery. On what dates should Grandma make appointments to see her doctor?

6 Daniel's son William is unwell. The doctor told Daniel to give William 8 mL of medicine 3 times per day for 5 days. The bottle contains 120 mL and Daniel gave William the first dose at 2 p.m. Will there be enough medicine for 5 days in one bottle?

3.03 Problem-solving strategies

Good problem-solvers use a variety of strategies. Here are 6 useful strategies, but there are plenty more.

- Draw a picture, diagram or make a model

- Make a table or list

- Guess, check and improve

- Work backwards

- Solve a simpler problem

- Break it into smaller pieces

We will use different problem-solving techniques as we work through Exercise 3.03.

Exercise 3.03 Problem-solving strategies

1 Jules has an appointment with the doctor on 26 April. Before the appointment she needs to have 2 blood tests: one 2 days before and the other 8 days before the appointment. On what dates does Jules need the blood tests?

   a Copy this timeline and mark the date of Jules' appointment on it.

   b Count the correct number of days backwards from Jules' appointment date to find the dates she needs to have the blood tests. What are the dates?

   c Which of the 6 problem-solving strategies did you use to solve this problem?

2 Year 1 students are going on an excursion. Adult tickets cost $25 and children's tickets cost $6. The school secretary ordered 20 tickets with a total cost of $196. How many adults are going on the excursion?

   a Guess the number of adults. If 2 adults have tickets, how many children have tickets? Calculate the total cost of the tickets. Is it correct?

   b Guess that there were 5 adults, calculate the number of children and the total cost of the tickets. Repeat until the total cost is $196.

   c Which of the 6 problem-solving strategies did you use to solve this problem?

3 Caroline uses dog treats to train her new puppy to go to the toilet outside. On the first day, she gave the puppy 32 dog treats and on each following day she gave the puppy half the number of treats than the previous day. The puppy was toilet trained on day 5, so she stopped the treats. How many dog treats did Caroline give her puppy?

Shutterstock.com/MAXXSIPHOTO

a How many treats did Caroline give the puppy on the second day?

b Copy and complete this table and calculate the total number of treats Caroline gave her puppy.

Day	1	2	3	4	5
Number of treats	32				

c Which of the 6 problem-solving strategies did you use to solve this problem?

4 Think about these 2 problems:

- A rocket travels 11.9 km every second. How far will it travel in 15.75 seconds?
- Jim walks 3 metres every second. How far does he walk in 4 seconds?

a Use this diagram to help you calculate how far Jim walks in 4 seconds. What mathematical operation using 3 and 4 gives the answer?

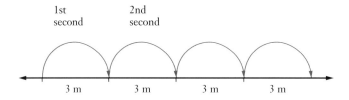

1st second   2nd second

3 m   3 m   3 m   3 m

b Use the same operation with 11.9 km and 15.75 seconds to calculate how far the rocket will travel.

c Which of the 6 problem-solving strategies did you use to solve this problem?

5 Every 50 km along the highway there is a rest area. Every 80 km there is a food stop and every 100 km there's a petrol station. How far from town is there a rest area, food stop and petrol station at the same place?

Town	R	R	R	R	R	R	R	R	R	R	R	R
km	50	100	150	200	250	300	350	400	450	500	550	600

a *R* on the number line above represents the locations of rest areas. Copy the number line and show the positions of food stops and petrol stations.

b Solve the problem.

c Which problem-solving strategy did you use to solve this problem?

6 Damien sells real estate. He receives 2% commission on the first $500 000 of the property he sells and 2.5% on any amount above $500 000. How much should Damien charge for his commission on a $960 000 sale?

a To solve this question, we need to separate the sale into $500 000 and the amount over $500 000. How much of the sale is over $500 000?

b Calculate the 2 separate parts of the commission, then the total commission.

c What problem-solving technique did we use to solve this problem?

3.04 Word clues

Sometimes, a problem contains words that can give us a clue or tip-off about the mathematical operations that we need to use to solve it. The table below shows lists of words that suggest when we have to add, subtract, multiply or divide.

Adding words (+)	Subtracting words (−)	Multiplying words (×)	Dividing words (÷)
total	more	lots of	share
sum	difference	product	how many times
increase	less	by	out of
combined	decrease	times	split
perimeter	deduct	groups of	parts
	change	per	at this rate
	how many left	at this rate	
	fewer	double (× 2)	
	reduce	triple (× 3)	
	remains	area	
	discount	volume	
	words that end in 'er' like bigger, slower	percentage of	

Be careful! Occasionally, words can have more than one meaning. Always check that your answer makes sense!

EXAMPLE 4

Let's revisit Kirsty's problem from Example 3:

'When Kirsty weighed herself, the scales showed 65 kg. She went on an unhealthy crash diet and lost 12% of her body weight. Calculate Kirsty's weight after her diet.'

What word clues can we find to help us decide whether to add, subtract, multiply or divide?

Solution

In this problem, '**lost**' means '**reduced**'. 'Reduced' is in the **subtract** column.

'**12% of**' is a '**percentage of**', which is in the **multiply** column.

To solve this problem, we will have to multiply and subtract.

a Identify the word clues in the following problem:

Sarah has a small flock of sheep to grow wool to use in her craft work. During lambing the number of animals increased from 12 to 15. How many lambs did Sarah get?

b Describe how you know that even though the word clue is in the 'addition' column, addition is NOT what we have to do to solve the problem.

Solution

a The word clue is 'increased'.

b In this problem the 'ed' at the end of 'increased' means the increase has happened already. We don't have to add, we have to undo an addition.
We have to **subtract**.

Exercise 3.04 Word clues

In each question identify the word clues and write whether we have to add, subtract, multiply or divide to solve the problem. You do *NOT* have to solve the problem.

1 Five people equally share the $2 million first prize in Lotto. How much will each person get?

Example 4

2 On Saturday, 6 friends are going to the movies. It's going to cost $26 per ticket. How much will the tickets cost?

3 Nick earns $780 per week. Each week his employer deducts $129 from Nick's earnings for tax. How much of Nick's wage is left after his employer deducts the tax?

4 Josie is a doctor in a medical centre. On average, she spends 15 minutes with each patient. If she decreases the average time she spends with each patient to 10 minutes, how many more patients can she see in an hour?

5 Newsagents sell 35 600 copies of the magazine 'Girlfriend' per month. How many copies of the magazine do newsagents sell each year?

6 The speed limit on the highway is 70 km/h, but during school zone times the speed limit is reduced to 40 km/h. How much slower do cars have to travel during school zone times?

7 Mariah placed an online order for 3 boxes of envelopes, 5 boxes of papers and a box of pens. Calculate the total number of boxes Mariah ordered.

8 'At this rate' is in both the multiplying and dividing columns. Which meaning of the clue 'at this rate' do we need to solve the following problem?

Troy assembles components in a factory. He assembles 12 components in an hour. At this rate, how long will it take him to assemble 72 components?

9 The gym instructor arranged the people in the class into 4 groups each containing 6 people. How many people are in the class?

In Questions 10 to 12, identify the word clues that have a different meaning from the ones in the table on page 60. What is the same about each of these questions?

10 Grandad died and left his cash to be shared equally among his 5 grandchildren. Each grandchild's share was $35 000. How much cash did Grandad have when he died?

11 Marco has been practising his golf putting. Now he can sink 6 out of 10 difficult putts, which is double the number he could sink before he started practising. How many difficult putts could he sink before he started practising?

12 Vicki received a $10 student discount when she bought some travel books from the bookshop. She only had to pay $65. How much would the books have cost without the student discount?

13 Go back to the questions and your answers in Exercise 3.02. Identify any word clues in each question and record what the clue is telling you to do. Keep your answers to use later.

ISBN 9780170412650

3.05 The 4 stages of problem solving

We can break down the process of solving a problem into 4 stages.

The 4 stages of problem-solving

Stage 1: WHAT is the problem? What do we know?

Stage 2: SOLVE the problem

Stage 3: CHECK the solution

Stage 4: PRESENT the solution

To demonstrate, we will revisit Kirsty's problem:

'When Kirsty weighed herself, the scales showed 65 kg. She went on an unhealthy crash diet and lost 12% of her body weight. Calculate Kirsty's weight after her diet.'

WHAT?

STAGE 1: WHAT IS THE PROBLEM? WHAT DO WE KNOW?

To find Kirsty's weight after her diet.

We know her weight before the diet was 65 kg and that she lost 12%.

SOLVE

STAGE 2: SOLVE THE PROBLEM

There's 2 word clues in this problem:

$$12\% \rightarrow \text{multiply, 'lost'} \rightarrow \text{subtract}$$
$$12\% \times 65 = 7.8, \text{ she lost 7.8 kg}$$
$$65 - 7.8 = 57.2 \text{ kg, her new weight is 57.2 kg}$$

[If you can't work out the solution, go back to Stage 1]

CHECK

STAGE 3: CHECK THE SOLUTION

Does it make sense? Have you solved the problem?

57.2 kg is less than 65 kg, the answer sounds correct and there's nothing else to work out.

[If your answer doesn't solve the problem, go back to Stages 1 and 2]

PRESENT

STAGE 4: PRESENT THE SOLUTION

Write the conclusion:

Kirsty weighed 57.2 kg after the diet.

Exercise 3.05 The 4 stages of problem-solving

Use the 4 stages of problem-solving to solve the problem given in Part A below. Having your answers to Exercise 3.02 and Question 13 of Exercise 3.04 will save you time and effort with the questions in Part B.

Part A

Saskia's horse weighs 500 kg and is feeding a foal. The horse's food is a mix of concentrate and paddock grass. Horses require 2% of their body weight in food each day. A mare (female horse) feeding a foal requires needs 70% of her food to be concentrate and the remainder paddock grass. How many kilograms of concentrate and paddock grass should Saskia give her horse each day?

Answer the following questions to help you learn what you have to do in each stage of solving a problem.

1   What do we have to work out and what do we know?

WHAT?

2   Are there any word clues that can help us know what to do?

3   What calculations do we need to do?

4   Complete the working out.

SOLVE

5   Check that the answer makes sense and comment about it.

CHECK

6   Complete the solution by writing the answer or conclusion.

PRESENT

Part B

For each question in Exercise 3.02, use the problem-solving stages to solve each problem. Remember that you identified the word clues in Question 13 of Exercise 3.04.

PROBLEM-SOLVING STAGES

Descriptions of the stages are listed below but they are NOT in order. Copy the blank grid and write the descriptions of each stage in the appropriate place. Ask your teacher to check your grid.

Checking our answer is very important. If it doesn't make sense we have to go back to stage 1 and start again.

In this stage, we write our conclusions and the solution to the problem.

We decide which problem-solving strategy to use. Sometimes we'll use more than one strategy. We look at the language in the problem for clues for what to do and then we solve the problem.

In this stage, we decide what the problem is asking us to find.

We read the problem carefully and we list all the information that is given.

WHAT?

SOLVE

CHECK

PRESENT

SOLUTION ^{TO} THE CHAPTER PROBLEM

SOLUTION TO THE CHAPTER PROBLEM

Problem

Caroline, Sandra and Sue are entering a craft competition and they have decided to share the costs equally. Sandra paid $102.35 for the material, Sue paid $67.20 for the equipment and Caroline paid $20 for the postage. Who should pay whom and how much should they pay?

Solution

WHAT?

STAGE 1: WHAT IS THE PROBLEM? WHAT DO WE KNOW?

To find the amount each person pays and to whom they pay.

Sandra paid $102.35, Sue paid $67.20, Caroline paid $20 and they will share the cost equally.

SOLVE

STAGE 2: SOLVE THE PROBLEM

Word clues: Share equally → divide

What strategy? Break it into smaller pieces.

$102.35 + $67.20 + $20.00 = $189.55

$189.55 ÷ 3 ≈ $63.18

Sandra and Sue have each paid too much. They require refunds from Caroline.

Sandra's refund = $102.35 − $63.18 = $39.17

Sue's refund = $67.20 − $63.18 = $4.02

CHECK

STAGE 3: CHECK THE SOLUTION

Caroline will pay $20 + $39.17 + $4.02 = $63.19.

That's right.

PRESENT

STAGE 4: PRESENT THE SOLUTION

Caroline should pay Sandra $39.17 and Sue $4.02.

Solving problems

Exercise 3.01

1 *Without* working out the answer, decide whether the answers given are wrong or could be right. Justify your answers.

a Phoebe is an electrician. She buys electrical cable in 100 m rolls that cost $31.80 each. She estimates that she will need 860 m of cable to wire a block of townhouses. How much will the cable cost? Answer: $94.60

b The price of a pair of shoes without GST included is $64. Calculate the price including 10% GST. Answer: $58.20

c When Skye pays her electricity bill early, she receives a 10% discount. Skye's electricity bill is $380. How much will she save if she pays the bill early? Answer: $342

d Billy is going to buy a motorbike. He will pay a $450 deposit and 12 monthly repayments of $180. How much will Billy pay for the bike? Answer: $2610

Exercise 3.02

2 Read the following problem but don't work out the answer.

Shelly has a new roll of wire 20 m long that she is going to use to wire up a new sound system in her unit. She needs 2 pieces of wire, one 8 m long and the other 3.6 m long. After Shelly is finished, how much wire will be left on the roll?

a What is the question asking?

b What information is included in the problem?

Alamy Stock Photo/YAY Media AS

3 For the following problem:

a list the information provided

b state what you have to work out.

Problem

| What does 'on the hour' mean? |

You are catching a bus to a football match at the stadium. Buses leave every hour on the hour and it takes 1 hour 20 minutes to get there. It takes you 30 minutes to get to the bus from your home. The match starts at 3:45 p.m. What is the latest time you can leave home?

4 In each problem, identify the word clues. You *don't* have to solve the problem.

a Ann and Mike are cleaning their neighbour's house. Ann worked for 3 hours and Mike worked for 1 hour. Their neighbour paid them $100 for the job. They want to share the money fairly. How much should they each receive?

b The bus fare to the mall is $3.40 one way or $4.50 return. How much cheaper is it to buy a return ticket than 2 one-way tickets?

c Coffee costs $4.20 per mug. Five friends each had a mug of coffee. Calculate the total cost.

d Harry receives a 15% discount when he buys car products because he is a member of a vintage car club. How much will Harry pay for car-cleaning products that normally cost $36?

5 Use the problem-solving stages to answer each question.

a A bike track is 500 m long. How many times do you need to ride around the track to ride 6 km?

b Parking at the beach costs 50c for 15 minutes. At 11:30 a.m. Max put $5 in the parking meter. At what time will his parking meter run out?

4.

APPLYING RATES

Chapter problem

Suzie needs to buy some concrete sealer to seal the bricks in her new garden wall. She needs 24 litres and the sealer is available in 4 L and 6 L containers.

What is the cheapest way for Suzie to buy the sealer?

WHAT WILL WE DO IN THIS CHAPTER?

- Identify and simplify rates
- Convert rates from one pair of units to another, for example, m/s to km/h
- Solve practical problems using rates, including speed, fuel consumption and unit pricing

HOW ARE WE EVER GOING TO USE THIS?

- Comparing prices of different brands of groceries to select the cheapest
- Calculating how long a trip will take if driving at a certain speed
- Determining quantities required for a painting or tiling job
- Calculating the quantity of fuel a car will use on a trip

4.01 Rates

Rates compare 2 different types of quantities.

The most common rate we use is **speed**, which compares distance travelled over time. If a car has a speed of 100 km/h, it means that the car travels 100 km for *each* hour it travels.

Unlike ratios, rates have units. We express these as the units of the first quantity **per** unit of the second quantity. 'Per' means 'for each one'.

EXAMPLE 1

Write each situation as a rate.

a   Andrew runs 200 m in 25 seconds.

b   The fuel for Boris' cultivator costs $35 for 20 L.

Solution

a   The units for this rate will be metres per second, or m/s.
Divide the number of metres by the number of seconds.

$$\text{Rate} = \frac{200 \text{ metres}}{25 \text{ seconds}}$$

$$= 8 \text{ m/s}$$

b   The units will be dollars per litre.
Divide the number of dollars by the number of litres.

$$\text{Rate} = \frac{\$35}{20 \text{ litres}}$$

$$= \$1.75/\text{L}$$

Exercise 4.01 Rates

1 Write each situation as a rate.

    a   200 km travelled in 4 hours

    b   75 words typed in 3 minutes

    c   250 L used in 5 hours

    d $87.50 for 5 kilograms

    e   400 m in 55 seconds

    f $65 earned in $2\frac{1}{2}$ hours

    g   500 g in 2 litres

    h   1800 revolutions in 3 minutes

2 Marcella uses 90 litres of composting manure on her flower bed which has an area of 5 m². What is this as a rate in L/m²?

3 Goran used his sprinklers to water his garden for $2\frac{1}{2}$ hours. He used 3375 litres of water. What is his rate of water use in L/h?

4 Write each situation as a rate.

a 320 L in 8 containers

b 10 litres consumed in $2\frac{1}{2}$ days

c 80 g in 15 cm³

d \$22.40 to hire 7 DVDs

e 175 mm rain in 5 days

f 290 km driven using 20 L of petrol

g \$20.52 for 1.8 m of curtain material

h 768 vibrations in 3 seconds

i 140 sheep on 3.5 hectares of land

j \$28.44 for 18 L of petrol

5 Dakota used 8 L of stone sealer to seal 72 m² of slate. Calculate this rate in m²/L.

6 John is preparing his budget for his holiday. For his 10-day holiday he is budgeting \$1200 for meals. Calculate John's meal budget in \$/day.

7 Marcus runs a barbecue for his sports club's market day. He has purchased the food shown in the table. Calculate the cost rate, including units, for each item.

Item	Cost and amount	Cost rate
Sausages	\$57.80 for 6.8 kg	a
Steak	\$93.60 for 7.5 kg	b
Onions	\$9.80 for 4 kg	c
Rolls	\$132 for 240 rolls	d
Tomato sauce	\$19.60 for 8 bottles	e

4.02 Converting rates

Sometimes, it is useful to be able to convert rates from one set of units to another. Doing this helps us to compare rates for different circumstances.

Converting rates from one set of units to another

- Write the rate as a fraction, including its units
- Convert the units on the numerator of the fraction to the new unit
- Convert the units on the denominator of the fraction to the new unit
- Simplify the rate.

WS

Converting rates

WS

More rates

Converting
rates

EXAMPLE 2

Convert 90 km/h to metres/second (m/s).

Solution

Convert 90 km into metres and 1 hour into seconds.

$$90 \text{ km} = 90 \times 1000 \text{ metres}$$
$$= 90\,000 \text{ m}$$
$$1 \text{ h} = 60 \text{ minutes}$$
$$= 60 \times 60 \text{ seconds}$$
$$= 3600 \text{ s}$$

Divide the number of metres by the number of seconds.

$$90 \text{ km/h} = \frac{90 \text{ km}}{1 \text{ h}}$$
$$= \frac{90\,000 \text{ m}}{3600 \text{ s}}$$

Simplify the rate.

$$= 25 \text{ m/s}$$

EXAMPLE 3

Convert 2.5 g/m^2 to kg/ha.

Solution

Change 2.5 g to kg and 1 m^2 to hectares.

There are 10 000 m^2 in 1 hectare.

$$2.5 \text{ g} = 2.5 \div 1000 \text{ kg}$$
$$= 0.0025 \text{ kg}$$
$$1 \text{ m}^2 = 1 \div 10\,000$$
$$= 0.0001 \text{ ha}$$

Simplify the rate.

$$2.5 \text{ g/m}^2 = \frac{2.5 \text{ g}}{1 \text{ m}^2}$$
$$= \frac{0.0025 \text{ kg}}{0.0001 \text{ ha}}$$
$$= 25 \text{ kg/ha}$$

Exercise 4.02 Converting rates

In this exercise, round answers to one decimal place where needed.

Example
2

1 Convert each speed into m/s.

    a   20 km/h   **b**   50 km/h   **c**   80 km/h   **d**   110 km/h

NELSON QMATHS 11. Essential Mathematics   ISBN 9780170412650

2 Alice is walking at a rate of 80 metres per minute. What is her speed in:

a km/h?   **b** m/s?

3 Convert each speed to km/h.

a 12.5 m/s   **b** 2400 m/minute   **c** 3500 m/h

4 Joshua fertilises his land at the rate of 4 g/m^2. Express this rate in kg/hectare.

Example **3**

5 A tap leaks at a rate of 30 mL per minute. What is this rate in litres per hour?

6 The steel mesh used in fencing weighs 1.2 kg/m.

a Convert this rate to g/mm.

b Convert this rate to tonnes/km.

c What do you notice about the answers to parts **a** and **b**?

d What is the rate in g/cm?

7 Jesse pumps water from his dam to irrigate his market gardens. The pump delivers water at a rate of 3.5 kL/hour. Express this rate as:

a L/h   **b** L/min   **c** L/s.

8 A 5 L container of varnish costs $14.

a What is the cost in $/L?

b Express this as a cost in cents/mL.

9 Jamaica's Usain Bolt set the world record for the 100 metre sprint. He ran the distance in 9.58 seconds.

a Express this as a rate in m/s (to one decimal place).

b Convert this speed to km/h (to 2 decimal places).

10 Sophie is painting her house before attempting to sell it. She uses 2 L of decking oil to paint her 38 m^2 wooden decks and 350 mL of black paint to paint the outside surface of her garden pots, an area of 4.6 m^2. Which surface, the wood deck or the garden pots, uses the larger rate of paint?

4.03 Unit pricing

The government requires supermarkets to display the **unit price** for all items. The unit price makes comparing the prices of different-sized packaging easy.

Alamy Stock Photo/RosaIreneBetancourt 5

Jonathan is calculating the unit price for rolls of paper towels. He has 15 m rolls for $1.65 each and 20 m rolls at $2.40 per roll. He decided to use 1 metre as the comparison unit.

a Calculate the unit price for each roll.

b Which roll is the better value for money?

Solution

a For the 15 m roll, calculate the cost of 1 metre by dividing by 15.

For the 15 m roll:
Unit price = $1.65 ÷ 15
= $0.11/m

For the 20 m roll, calculate the cost of 1 metre by dividing by 20.

For the 20 m roll:
Unit price = $2.40 ÷ 20
= $0.12/m

b The better price is the lower unit price.

The 15 m roll is the better value.

EXAMPLE 5

The same cooking oil is available in two sizes of bottle:

500 mL for $4.75 and 750 mL for $6.45.

Unit pricing

> A 'unit' varies depending on the items being compared. It can be any convenient unit; for example, 100 g, 1 L, 100 mL or 1 m.

a Calculate the unit price for each bottle.

b Which size of bottle is the better value?

Solution

a For each bottle, calculate the price for 100 mL.
500 mL is 5×100 mL
Divide the price of the 500 mL bottle by 5.
750 mL is 7.5×100 mL
Divide the price of the 750 mL bottle by 7.5.

Unit price of the 500 mL bottle is:
$4.75 ÷ 5 = 0.95 per 100 mL

Unit price of the 750 mL bottle is:
$6.45 ÷ 7.5 = 0.86 per 100 mL

b The smaller unit price is the better value.

The 750 mL bottle is the better value.

Exercise 4.03 Unit pricing

1 Calculate the price of 1 m of wallpaper for each size roll.

Example
4

  a 5 m for $11.50

  b 10 m for $22.00

  c 20 m for $45.00

2 The same seafood sauce is available in 50 mL bottles for $3.60 and 80 mL bottles for $5.60.

  a For each size of bottle, calculate the price per 10 mL.

  b Which size of bottle is the better value?

3 A box containing 1 dozen eggs costs $4.80 and a box containing 6 eggs is $2.10.

  a Calculate the price per egg in each box.

  b Which size box is the better value?

  c Damien is buying 18 eggs. What is the cheapest way for him to do it?

4 Supermarket signs show the packet price and the unit price for 3 packets of breakfast cereal. Which size packet is the best value for money? Justify your answer.

450 g for $4.49
Unit price $1.00 per 100 g

350 g for $3.00
Unit price $0.86 per 100 g

330 g for $4.25
Unit price $1.29 per 100 g

5 Rice is available in 3 different-sized packets:

- 1 kg for $3.50

- 2 kg for $8.00

- 5 kg for $17.99

  a What would be a sensible unit to use for a price comparison?

  b Determine which size packet is the best value.

6 The yellow spice, saffron, is the world's most expensive spice.

The price of 1 g of saffron is 3 times the price of 1 g of 24 carat gold.

Carmen is ordering some saffron online.

2 g for $290

2.5 g for $350

5 g for $695

7.5 g for $1065

  a Calculate the unit (1 g) price for each size.

  b Carmen is going to order 7.5 g. What is the cheapest way for her to do it?

7 Breakfast biscuit cereal QBix are on special because the 'use before date' is only 5 weeks away. A box containing 24 biscuits is priced at $4.56 and a box containing 48 biscuits is $8.64.

Mae likes to have one biscuit for breakfast on most days. Which size box do you recommend she buy?

8 Hair shampoo is available in 2 sizes of bottle:
- 750 mL for $8.95
- 1 L for $11.50

Which size bottle is the better value? Justify your answer.

Chapter problem

You've covered the skills required to solve the chapter problem. Can you solve it now?

4.04 Rate problems

Rate problems

To solve problems involving rates, we either **multiply** or **divide** by the rate:
- Identify the rate and express the units of the rate as a fraction (e.g. $\frac{km}{h}$).
- To find the numerator of the fraction, multiply by the rate.
- To find the denominator of the fraction, divide by the rate.

Rapid rates

Rate skills

EXAMPLE 6

Noah hires a small car for $42 per day. How much will the hire cost be if Noah keeps the car for 16 days?

Rate problems

Solution

The rate is $42/day. We can express the units as $\frac{\$}{day}$. To find the number of dollars, which is the numerator of the fraction, multiply by the rate.

Hire cost = 16 days × $42/day
= $672

Check that the answer is realistic, then write the answer.

It will cost Noah $672 to hire the car for 16 days.

EXAMPLE 7

Amy sends the fruit she grows on her farm to market in cartons that hold 12 kg of fruit per carton. How many cartons will she need to pack 180 kg of fruit?

Solution

The rate is 12 kg/carton. The units are $\dfrac{\text{kg}}{\text{carton}}$.

To find the number of cartons, which is the denominator, divide by the rate.

Number of cartons
$= 180 \text{ kg} \div 12 \text{ kg/carton}$
$= 15 \text{ cartons}$

Check and write the answer.

Amy will need 15 cartons.

Exercise 4.04 Rate problems

1 Lizzie plants tomato seedlings in rows at a rate of 28 seedlings per row. How many seedlings are there in:

   a   4 rows?   **b**   7 rows?   **c**   12 rows?

2 Jayden earns $12.40/h as a barista in a cafe. How much is he paid for working 20 hours?

3 Kane proudly claims that he can hike at an average rate of 5 km/h even in difficult conditions. At this rate, how long will it take him to hike:

   a   15 km?   **b** $7\frac{1}{2}$ km?   **c**   24 km?

4 Mandarins cost $3.20/kg. How many kilograms of mandarins can Peter buy for $20?

5 Danielle is growing nectarine trees. She can expect a yield of 80 kg of fruit per tree.

   a   How many kilograms of nectarines can Danielle expect to produce if she plants 9 trees?

   b   How many trees would Danielle need to plant if she wanted to produce 1200 kg of nectarines?

6 A plane flies at a speed of 840 km/h.

   a   How far does the plane travel in $6\frac{1}{2}$ hours?

   b   Calculate how long it would take the plane to fly from Melbourne to each city, given their distances. Answer in hours correct to one decimal place.

      i   Brisbane: 1670 km

      ii   Auckland, New Zealand: 2626 km

      iii   Perth: 3430 km

      iv   Los Angeles, USA: 12 773 km

> We will solve more speed problems in Chapter 16, using a formula for speed.

7 The yields Maxine achieves from different vegetable crops are shown below.

Crop	Yield (vegetables/m^2)
Beetroot	40
Lettuce	18
Cabbage	8

a How many beetroots can Maxine expect to get from her 24 m^2 beetroot plot?

b How many lettuces will she grow in her 15 m^2 plot?

c What area does she need to grow 450 lettuces?

d What area does she need to grow 760 cabbages?

8 Aaron uses fertiliser on his farm at a rate of 24 kg/hectare. How many kilograms of fertiliser does he require to cover 32 hectares?

9 A dripping tap leaks water at the rate of 25 mL/minute.

a How much water will leak from the tap in 30 minutes?

b How much water will be lost in 1 hour?

> Remember, there are 60 minutes in an hour.

c Convert your answer to part **b** to litres.

d How many litres will be lost in 24 hours?

e How many minutes will it take for the tap to leak 1 litre of water?

> Remember, there are 1000 mL in 1 L.

f How many hours will it take for the tap to leak 60 L of water?

10 Vijay is saving to purchase a car. He can save $400 per week.

a How much can Vijay save in 20 weeks?

b The car he wants costs $14 000. How long will it take Vijay to save this amount?

c A newer car is available but it costs $15 200. How much longer would it take Vijay to save for this car?

11 Chloe earns $24.50/h as a secretary. How many hours will she need to work to earn:

a $441?   **b** $857.50?   **c** $588?

12 Elise wants to hang curtains on 12 windows in her house. Each window requires 1.8 metres of material, which costs $12.50 per metre.

    a How many metres of material will Elise need for all 12 windows?

    b How much will it cost?

    c Elise has budgeted only $225 for the material. How many metres of material can she buy if she sticks to this budget?

    d How many windows can she curtain in this situation?

13 Jacob uses a small aircraft to do crop dusting. His plane uses fuel at a rate of 28 litres per hour. In one week, Jacob flew his plane for 18 hours.

    a How many litres of fuel did Jacob use that week?

    b That week, aviation fuel cost $2.25/L. How much was his fuel bill for the week?

    c Jacob charges $124/h when he is crop dusting. How much did he charge for the week crop dusting?

    d In a different week, Jacob had only 420 litres of fuel available. How many hours could he fly in that week?

4.05 Fuel consumption

Fuel
consumption

How much
petrol?

The **fuel consumption** of a vehicle is the amount of fuel the vehicle uses per distance travelled. It depends on many factors, including the size and efficiency of the engine, the mass of the vehicle and whether it is driven on city or country roads.

Fuel consumption is a rate expressed as the number of litres used per 100 km travelled (L/100 km). The lower the rate, the less fuel the vehicle uses.

Fuel consumption

$$\text{Fuel consumption} = \frac{\text{fuel used in L}}{\text{distance travelled in km}} \times 100$$

For example, the fuel consumption for a Honda Jazz GLi is 5.8 L/100 km, and for the much larger V6 Holden Commodore it is 9.8 L/100 km.

There are 3 types of fuel consumption questions.

- What is the fuel consumption rate in L/100 km? Divide the fuel used by the distance travelled and multiply by 100.

- How much fuel will the car use? To find litres (L), multiply by the rate.

- How far can the car go? To find km travelled, divide by the rate.

EXAMPLE 8

The dashboard computer in Sue's car shows that over the last 5000 km travelled the car used fuel at a rate of 6.9 L/100 km.

a How much fuel did Sue's car use in the last 5000 km?

b At an average price of $1.55/L, how much did the petrol to travel 5000 km cost?

Solution

a Express 6.9 L/100 km as $\dfrac{6.9 \text{ L}}{100 \text{ km}}$

To find litres, multiply by the rate.

$$\text{Fuel consumption} = \dfrac{6.9 \text{ L}}{100 \text{ km}} = 0.069 \text{ L/km}$$
$$\text{Fuel used} = 5000 \text{ km} \times 0.069 \text{ L/km}$$
$$= 345 \text{ L}$$

b Express $1.55/L as $\dfrac{\$1.55}{\text{L}}$

To find cost ($), multiply by the rate.

$$\text{Cost} = 345 \text{ L} \times \$1.55/\text{L}$$
$$= \$534.75$$

Shutterstock.com/Art Konovalov

EXAMPLE 9

Jesse bought a second-hand car. This week, he travelled 425 km and used 32 L of fuel. Calculate correct to one decimal place the car's fuel consumption in L/100 km.

Solution

$$\text{Fuel consumption} = \dfrac{\text{fuel used in L}}{\text{distance travelled in km}} \times 100$$

$$\text{Fuel consumption} = 32 \text{ L} \div 425 \text{ km} \times 100$$
$$= 7.5294...$$
$$\approx 7.5 \text{ L/100 km.}$$

EXAMPLE 10

Saskia's BMW has a fuel consumption rate of 7.1 L/100 km. How far (to the nearest kilometre) can the car travel on 60 L of fuel?

Solution

Express 7.1 L/100 km as $\dfrac{7.1\text{ L}}{100\text{ km}}$

To find km, divide the fuel used by the rate.

Fuel consumption $= \dfrac{7.1\text{ L}}{100\text{ km}} = 0.071$ L/km

Distance $= 60$ L $\div 0.071$ L/km

$\qquad\quad = 845.0704...$

$\qquad\quad \approx 845$ km

Exercise 4.05 Fuel consumption

1 The petrol consumption rate of Jack's car is 5.9 L/100 km.

    a How many litres of petrol will the car use on a 850 km trip?

    b How much will the fuel cost for the trip if petrol costs $1.60/L?

2 Tia's old Holden used 25 L of fuel to travel 180 km. Calculate the car's fuel consumption in L/100 km.

3 Jeff's VW Golf uses diesel fuel. It has a fuel consumption rate of 6.1 L/100 km. How far can Jeff travel on 50 litres of diesel fuel?

4 Charles used 18 L of petrol to drive 190 km. Calculate his car's fuel consumption in L/100 km.

5 A motoring website tested the fuel consumption for a number of SUV vehicles by measuring the fuel consumed for the same 790 km journey. Calculate the fuel consumption of each vehicle correct to one decimal place.

	Vehicle	Litres of fuel used
a	Kia SUV	80.1
b	BMW	69.9
c	Toyota Rav4	67.2
d	Subaru Outback	57.7
e	Range Rover	61.6

6 Carol's second-hand Toyota Camry uses 7.5 L/100 km. The petrol tank holds 60 litres of fuel. How far can Carol travel on one tank of petrol?

7 To save fuel, the motor in Lisa's SUV turns itself off when the car stops in traffic. With this fuel-saving option turned on, the car uses 8.9 L/100 km, but with it turned off, it uses 10.0 L/100 km. Each week, Lisa travels 650 km.

   a How much less fuel per week does Lisa use with the fuel-saving option turned on?

   b At an average price of $1.50/L, how much does Lisa save by turning on the fuel-saving option?

8 Alistair's Lexus Sport uses premium fuel (98 petrol) with a fuel consumption of 11.1 litres per 100 km.

   a How much petrol will his car use on a 425 km trip to the vineyards? Answer correct to the nearest litre.

   b Calculate the cost of fuel for the journey at $1.75 per litre.

9 Josh lives on a large property. He uses a ride-on mower to cut the grass, but he's heard that ride-on mowers use much more fuel than a hand mower and he's concerned about the effect on the environment. He decided to investigate. He divided his biggest paddock into two equivalent sections and mowed one with a ride-on and the other with a hand mower. He summarised the results in a table.

Method	Fuel usage rate	Time taken
Hand mower	600 mL/h	3 hours
Ride-on mower	3.6 L/h	30 minutes

   a Calculate the actual amount of fuel Josh used to mow each section.

   b Which method of cutting the grass would you recommend Josh use? Why?

10 Tara's chainsaw uses fuel depending on the hardness of the wood she is cutting. When cutting large gum trees, it uses fuel at a rate of 310 mL in 45 minutes.

   a Calculate this fuel usage in L/h. Answer correct to one decimal place.

   b Tara estimates that it will take her 8 hours to cut up a large gum tree. How much fuel will Tara require?

   c Tara only has 3 L of fuel. How long will she be able to use the chainsaw?

CLUELESS CROSSWORD

Copy this crossword and position the keywords from this chapter in the crossword. Then write a set of clues for your crossword.

comparison   convert   fuel consumption   per

rate   speed   time   unit pricing

SOLUTION ^{TO}_{THE}
CHAPTER PROBLEM

Problem

Suzie needs to buy some concrete sealer to seal the bricks in her new garden wall.
She needs 24 litres and the sealer is available in 4 L and 6 L containers.

What is the cheapest way for Suzie to buy the sealer?

Solution

WHAT?

STAGE 1: WHAT IS THE PROBLEM? WHAT DO WE KNOW?

We have to work out the cheaper way to buy the sealer.

Cost of one 4 L container is $55.

Cost of one 6 L container is $87.

SOLVE

STAGE 2: SOLVE THE PROBLEM

Calculate the cost per litre for each container

Cost per litre in the 4 L container = $55 ÷ 4

= $13.75

Cost per litre in the 6 L container = $87 ÷ 6

= $14.50

It is cheaper to buy the sealer in 4 litre containers, so 6 × 4 L containers.

CHECK

STAGE 3: CHECK THE SOLUTION

Suzie needs 24 L.

If she buys 4 L containers, she will need 6.
If she buys 6 L containers, she will need 4.

Total cost of the sealer in 4 L containers = $6 \times \$55$

$= \$330$

Total cost of the sealer in 6 L containers = $4 \times \$87$

$= \$348$

PRESENT

STAGE 4: PRESENT THE SOLUTION

It is cheaper to buy 6×4 L containers of sealer.

4. CHAPTER REVIEW

Applying rates

1 An electrician charged $160 for $2\frac{1}{2}$ hours work. Calculate her rate in dollars/hour.

2 Convert a speed of 75 km/h into m/s. Answer to one decimal place.

3 The Earth travels at approximately 30 km/s in its orbit around the Sun.
What is this speed in km/h?

4 The same brand of olive oil is available in 3 sizes.

	Size	Cost
Small	375 mL	$5
Medium	750 mL	$8
Large	1 L	$13

   a   Calculate the cost per 100 mL in each size.

   b   Which size is the best value for money?

5 Denis charges $55/m^2 to lay marble tiles. How much will he charge for laying 42 m^2 of marble tiles?

6 A plumber charges a $50 callout fee which includes the first 15 minutes. For calls longer than 15 minutes, he charges an additional $28 per 15 minutes or part thereof.
How much will the plumber charge for a callout that takes 45 minutes?

7 Olivia has been offered a new job. She has to choose her pay rate:
$36.50 per hour or a flat rate of $1400 per week.
In the job, Olivia will be working 40 hours per week. Which pay rate is the better deal?

8 Kim is comparing the fuel consumption of her 2 cars.

Car	Distance travelled	Fuel used
White car	620 km	53 L
Red car	470 km	43 L

   a   Calculate the rate at which both cars used fuel in L/100 km. Answer to one decimal place.

   b   Which car is the more fuel-efficient?

5.

SHOW ME THE GRAPH

Chapter problem

During a drought, the water authority produced a graph to show the decreasing supply of water in a dam due to the water usage of local residents. Is the decrease in water supply as bad as the graph shows? If the graph gives a false impression, identify what has been done to create this impression.

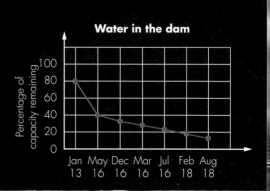

Water in the dam

WHAT WILL WE DO IN THIS CHAPTER?

- Interpret information presented in a variety of graphs and two-way tables: picture graphs, column graphs, line graphs, conversion graphs, step graphs and pie graphs
- Determine which type of graph is best suited to display a particular set of data
- Use spreadsheets to graph data
- Identify misleading graphs appearing in the media and in advertising

HOW ARE WE EVER GOING TO USE THIS?

- When reading a graph that is presented in a newspaper, magazine or website
- When drawing a graph to illustrate data to others
- When noticing a graph is being misleading in a report or advertisement

Exercise 5.01 Interpreting graphs

1 **Picture graphs** use symbols to represent quantities. This picture graph shows the number of cars passing a local high school at different times of day.

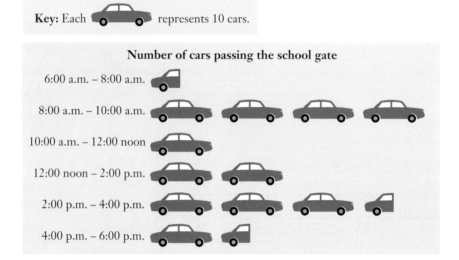

a What does 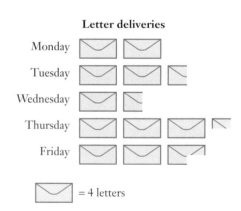 represent?

b How many cars passed the school between 2 p.m. and 4 p.m.?

c What time of day had the least traffic?

d Write a paragraph describing the traffic flow and suggesting reasons for the differences.

2 This picture graph shows the number of letters delivered to an office each day last week.

a How many letters were delivered on Thursday?

b On which day were 11 letters delivered?

c How many letters were delivered to the office last week?

d What is one disadvantage of this type of graph?

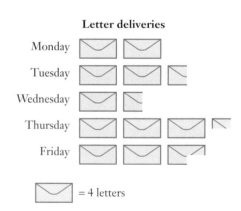

3 **Column graphs** are mostly used for data that is in categories. This column graph shows the populations of the 8 Australian state capitals in 2016.

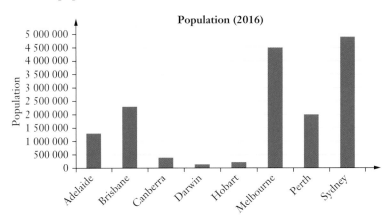

a What does one interval on the vertical axis represent?

b Which city has the largest population? Estimate its population from the graph.

c Which city has the smallest population? Estimate its population from the graph.

d Explain why we can only *estimate* the population from the graph.

e Which city has an approximate population of 1.3 million?

f Use the Internet to find the current populations of each city.

4

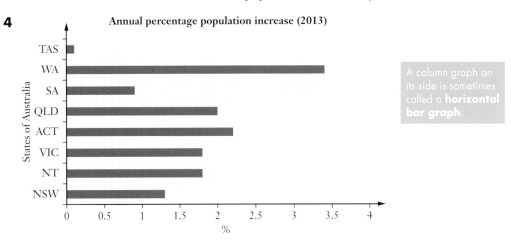

A column graph on its side is sometimes called a **horizontal bar graph**.

a By what percentage did Queensland's population increase in 2013?

b In which state was the percentage population increase 1.3%?

c Which 2 states had the same percentage increase?

d Does this graph tell you anything about the actual population of each of the states? Explain your answer.

e The population of Tasmania in 2012 was approximately 512 400. Calculate the actual increase in Tasmania's population.

5 We use a **clustered column graph** for categories we want to compare. This graph shows the nutritional information for different foods.

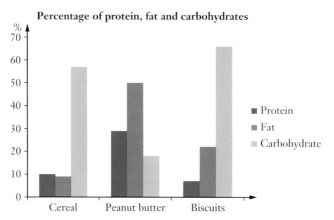

Percentage of protein, fat and carbohydrates

a Which food has the highest percentage of protein?

b Which food has the highest percentage of carbohydrate?

c Which food has the lowest percentage of fat?

d What is the difference in the percentage of fat in peanut butter and biscuits?

e If you were on a low carbohydrate diet, which food could you include in your diet?

f If you were on a low protein diet, which foods could be included in your diet?

g Jack had a 140g serving of cereal for his breakfast. Approximately how many grams of protein did the serving of cereal contain?

h Lizzie has to limit her carbohydrate intake. What foods should she avoid?

6 This clustered column graph shows the number of motor vehicle thefts per year for 3 different areas.

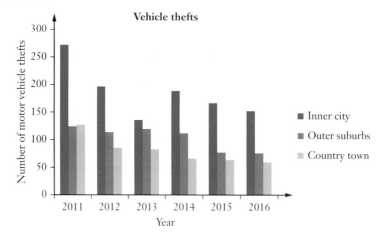

Vehicle thefts

a Describe the trend in motor vehicle thefts in the country town.

b In which year do the outer suburbs have the lowest number of thefts?

c In which year is the difference between the number of thefts in the inner city and the country town smallest? Estimate this difference.

d Write a brief paragraph describing the differences and similarities in the number of motor vehicle thefts in these 3 areas.

7 A **line graph** shows data that changes over time. The graph below shows the dramatic changes in home loan interest rates during the 1990s.

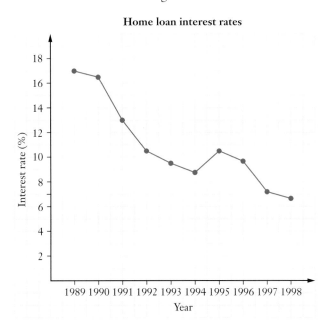

Home loan interest rates

a By how much did the interest rates drop between 1989 and 1998?

b Between which 2 years was the drop the greatest?

c Between which 2 years did interest rates increase?

d Anand borrowed $120 000 in 1989 to buy a house. If a flat rate of interest was used, how much interest did he pay in one year?

e Catriona borrowed the same amount in 1997.

  i Approximately how much interest did she pay in one year?

  ii How much more did Anand pay in one year?

8 This graph shows the electricity consumption for a household over a fortnight.

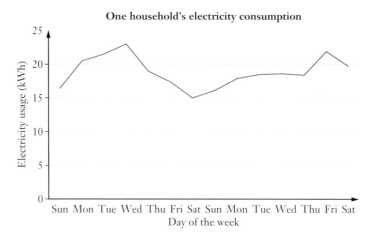

One household's electricity consumption

a What was the lowest daily rate of electricity usage?

b What was the highest daily rate of electricity usage?

c The usage is above 20 kWh for 4 days in the fortnight.
Which 4 days were these?

9 A **conversion graph** is used to convert from one unit to another. This graph shows conversions between degrees Fahrenheit and degrees Celsius.

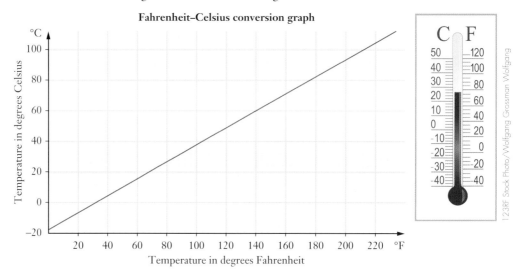

Fahrenheit–Celsius conversion graph

a Convert a temperature of 68°F to °C.

b Water boils at 100°C. What Fahrenheit temperature is this?

c Water freezes at 0°C. What Fahrenheit temperature is this?

d When Australian temperatures were measured in Fahrenheit, a day when the temperature reached 100°F, or a century, was considered a very hot day. What is this temperature in degrees Celsius?

NELSON QMATHS 11. Essential Mathematics

ISBN 9780170412650

10 This conversion graph can be used to convert Australian dollars to euros (€), the currency used in Europe.

Currency conversion graph

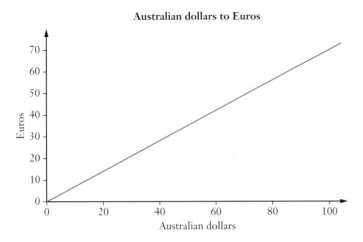

a Convert $15 to euros.

b Convert €50 to Australian dollars.

c A meal in Paris costs €35. What is this in Australian dollars?

d Gustav is visiting Australia. He has €25 left on his debit card. Is this enough to pay for a $25 meal at the local club?

e Calculate how many euros you would get for $220.

11 This conversion graph converts between calories and kilojoules, units of energy used in food nutrition.

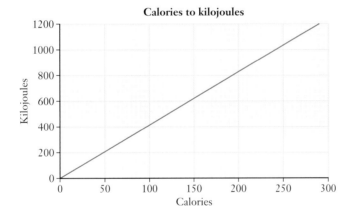

a Convert 100 calories to kilojoules.

b Nami's muesli bar contains 700 kJ of energy. How many calories is this?

c Jordan burned 500 calories working out at the gym. How many kilojoules is this?

d The average daily allowance for a healthy diet is 8700 kJ. How many calories is this?

12 A **step graph** is a line graph of
'broken' horizontal intervals that
look like steps.

Prem repairs air conditioners.
This step graph shows his charges
according to the hours of work.

a How much does Prem charge
for any time under an hour?

b How much does he charge for
exactly 2 hours?

c How much would Prem charge
for working at your house for
4.5 hours?

d How much does he charge for
each additional hour he stays to
make repairs?

e Why does he charge a lot more for the first hour
than for additional hours?

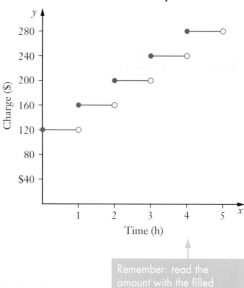

Air conditioner repair rates

Remember: read the
amount with the filled
circle, not the open circle.

13 This graph shows the charges for parking at an airport.

a How much does the airport charge
for 2 hours of parking?

b Anh parked for 3 hours and
23 minutes. How much did his
parking cost?

c When Madison went to the airport
to pick up her mum, the plane was
delayed. She had to park for
6 hours and 42 minutes. How
much did this cost?

d According to this graph, what is
the most you will have to pay for
parking at the airport?

e What is the average charge per hour for parking at the airport for 6 hours?

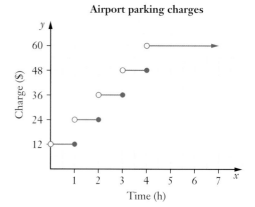

Airport parking charges

14 This **pie graph** shows the results of a survey about people's favourite sport to watch on TV.

  a What is the most popular sport?

  b What is the least popular sport?

  c Which 2 sports have sectors approximately the same size?

  d Rugby league was the favourite sport of 24% of the 3500 people surveyed. How many people preferred to watch rugby league on TV the most?

  e Do you think the results of a survey like this would be different if the survey was taken in different places? Justify your answer.

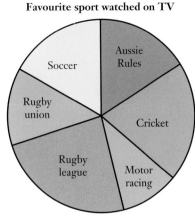

Favourite sport watched on TV

15 This graph shows the costs of producing a book as a percentage of the book price.

  a Which is the highest cost in producing a book?

  b List the costs in order from highest and lowest.

  c Royalties to authors account for 15% of the cost of production.
A new textbook costs $65.

    How much is the royalty payment?

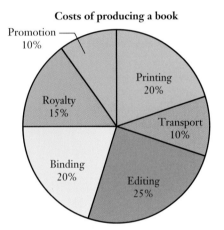

Costs of producing a book

16 NelsonNet Airlines surveyed passengers on a flight from Brisbane to Sydney on their reasons for travelling to Sydney.

  a What was the main reason people were travelling from Brisbane to Sydney?

  b Of the 176 passengers on this flight, how many were travelling for work/business?

  c This survey was conducted on a Friday evening. Would you expect the same results if the survey was conducted on a Monday evening? Justify your answer.

Reasons for travel

5.02 Two-way tables

Two-way tables show two characteristics of any set of data.

EXAMPLE 1

This two-way table shows the results of a survey on right/left-handedness and right/left-footedness.

	Right-footed	Left-footed	Total
Right-handed	50	10	60
Left-handed	12	8	20
Total	62	18	80

a What percentage of people in this survey are right-handed?

b What percentage of the right-handed people in this survey are left-footed? Answer correct to 1 decimal place.

> If you're not sure what number to put on the bottom of the fraction, look at the quantity following 'of' in the expression 'what percentage of …'.

Solution

a There are 80 people in the survey and 60 people are right-handed.

$$\text{Percentage} = \frac{60}{80} \times 100$$
$$= 75\%$$

b 60 people are right-handed and of those, 10 are left-footed.

$$\text{Percentage} = \frac{10}{60} \times 100$$
$$= 16.666...\%$$
$$\approx 116.7\%$$

Exercise 5.02 Two-way tables

1 Jarrod manages a muffin and scones shop. He surveyed locals on whether they preferred muffins or scones. The results are summarised in this table.

	Prefers muffins	Prefers scones	Total
People aged over 30	24	36	60
People aged 30 and younger	48	12	60
Total	72	48	120

a How many people did Jarrod survey?

b Which age group prefers scones to muffins?

c Overall, which item do Jarrod's customers prefer?

d How many of the people who prefer scones are over 30 years old?

e What percentage of people aged 30 or younger prefer muffins?

ISBN 9780170412650

2 This table shows the smoking habits of the adult population of Nelson Waters.

	Men	Women	Total
Smoker	3500	2400	i
Non-smoker	7500	8600	ii
Total	iii	iv	v

a Copy the table and complete the missing values in **i** to **v**.

b What is the adult population of Nelson Waters?

c What fraction of the adults smoke?

d What percentage of the adults are non-smokers? Answer to 2 decimal places.

e What percentage of the adult women who live in Nelson Waters are smokers? Answer to 1 decimal place.

3 Simone runs a café. He recorded data about who buys hamburgers and chicken wraps.

	Hamburgers	Chicken wraps	Total
Male	77	43	i
Female	31	64	ii
Total	iii	107	iv

a Calculate the missing totals.

b How many customers were counted?

c How many females bought chicken wraps?

d What percentage of customers bought hamburgers? Answer to 2 decimal places.

e What percentage of men bought chicken wraps? Answer to 1 decimal place.

f Simone is expecting a busload of men's football teams to visit the café. Should she prepare more hamburgers or chicken wraps? Justify your answer.

4 Nick found this table in a news article for the local gym. The data was obtained from surveying people from the local area about whether they exercise regularly.

	Men	Women	Total
Regular exercise	75	i	158
No regular exercise	iii	77	ii
Total	iv	v	345

a Copy the table and complete the missing values in **i** to **v**.

b What percentage of the people interviewed were women who exercised regularly? Answer correct to 2 decimal places.

c What percentage of the men interviewed have no regular exercise? Answer correct to 2 decimal places.

d If the gym was aiming to encourage people who don't exercise to join the gym, should they aim the advertisement at men or women? Justify your answer.

5 Alannah surveyed a group of people for their opinion on changing the Australian flag.

	Aged under 35	Aged 35 and over	Total
Change the flag	2570	1060	i
Keep the flag	6350	2780	ii
Total	iii	iv	v

a Copy the table and complete the missing values in **i** to **v**.

b What percentage of those surveyed wanted to keep the flag? Answer correct to 2 decimal places.

c What percentage of those under 35 wanted to change the flag? Answer correct to 2 decimal places.

d Based on this survey, should the Government change the Australian flag? Justify your answer.

6 Police are testing a new lie detector machine. The results are shown in this table.

	Machine judged as true	Machine judged as a lie	Total
True statements	65	15	i
False statements	ii	30	40
Total	iii	iv	v

a Copy the table and complete the missing values in **i** to **v**.

b On how many statements did the police test the machine?

c For how many statements did the machine correctly determine whether the statement was true or false?

d What percentage is this? Answer correct to 1 decimal place.

e What percentage of *true* statements were incorrectly identified by the machine? Answer correct to 1 decimal place.

ISBN 9780170412650

5.03 Everyday graphs

Exercise 5.03 Everyday graphs

1 This type of graph commonly appears on a phone bill.

a How much is the most recent bill?

b How does this compare with the same time last year?

c What is the lowest amount paid for one month?

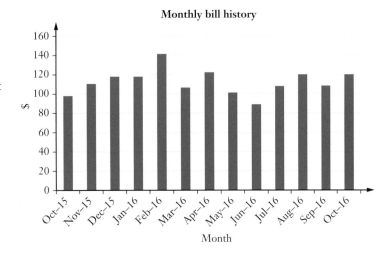

d Why would the telephone company include this graph on your telephone bill?

2 This is a typical electricity bill graph.

a How many months does each bill cover?

b What is the most recent amount paid?

c Give a reason why the amounts in June to August are much bigger than for the other months.

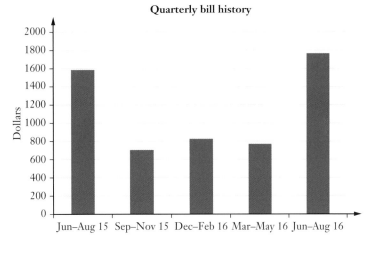

d Why is this type of graph useful to customers?

3 Gas bills usually show 'Your average daily gas usage' measured in megajoules (MJ).

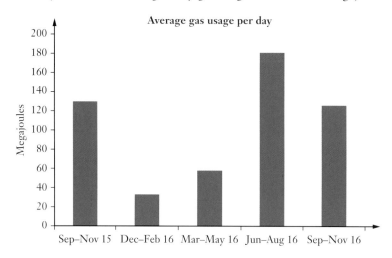

a How many months in total does this graph cover?

b What is the highest average daily gas usage?

c At what time of year is the most amount of gas used? Why might this occur at this time of year?

d Why is this type of graph useful to customers?

4 This table shows the interest rates offered by NQM Bank for term deposits over differing periods of time. With a term deposit, the money must stay in the account for the whole time and interest is paid at the end.

Minimum deposit	No of months.												
	1	2	3	4	5	6	7	8	9	10	11	12	24
$1000	2.05	2.05	2.05	2.05	2.05	2.05	2.05	2.05	2.05	2.05	2.05	2.05	2.05
$5000	2.05	2.05	2.55	2.30	2.30	2.75	2.30	2.30	2.30	2.30	2.30	2.75	3.15
$20 000	2.05	2.05	2.55	2.30	2.30	2.75	2.30	2.30	2.30	2.30	2.30	2.75	3.15
$50 000	2.05	2.05	2.55	2.30	2.30	2.75	2.30	2.30	2.30	2.30	2.30	2.75	3.15
$100 000	2.05	2.05	2.55	2.30	2.30	2.75	2.30	2.30	2.30	2.30	2.30	2.75	3.15

a What is the lowest interest rate offered?

b Regina has $5000 that she wants to spend on a holiday in 10 months. What is the best interest rate available to her?

c Stephen invests $20 000 for 12 months. How much interest does he earn?

d What are the disadvantages of putting your money in a term deposit for 24 months?

5 This graph shows the number of sunny days per month last year in Broome, Western Australia.

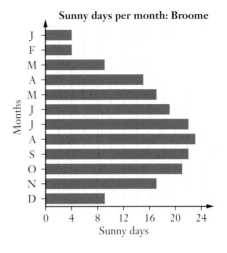

Sunny days per month: Broome

a Which month had the most sunny days?

b Which month had the least sunny days?

c List 3 pairs of months that had approximately the same number of sunny days.

d How many months of the year had 16 or more sunny days?

e Andrea is getting married in Broome. She would like a sunny day for her wedding. In which 3-month period should she plan her wedding to be most likely to get a sunny day?

f Give another reason why this graph might be useful.

6 A climate graph combines temperature and rainfall information. The climate graph below is for Emerald, west of Rockhampton.

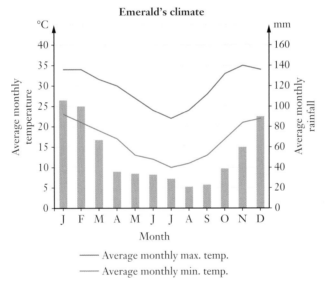

Emerald's climate

a What is the average maximum temperature in Emerald in December?

b Which 3 months have the coldest minimum temperatures?

c Describe Emerald's average monthly rainfall pattern.

ISBN 9780170412650

7 This is the climate graph for Toowoomba.

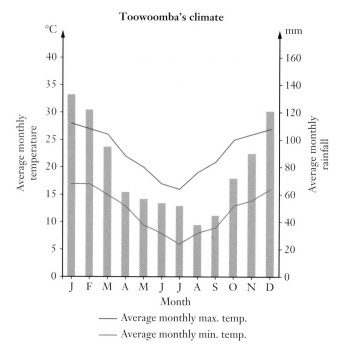

Toowoomba's climate

— Average monthly max. temp.
— Average monthly min. temp.

a Which 4 months have the lowest rainfall?

b What is the difference between the average maximum and minimum temperatures in Toowoomba in February?

c How much more rain falls during October than during July?

d Explain why it would be useful to have a graph like this for any place you might consider moving to.

8 Is Happy Valley in the northern hemisphere or the southern hemisphere? Explain your answer.

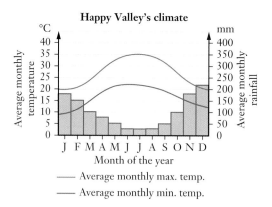

Happy Valley's climate

— Average monthly max. temp.
— Average monthly min. temp.

9 This picture graph was published by the Australian Red Cross Blood Service.

The red cells from your donation are being used in the following ways:

34% Cancer and blood diseases

19% Other causes of anaemia

18% Surgical patients including open heart surgery and burns

13% Other medical problems including heart, stomach and kidney disease

10% Orthopaedic patients including fractures and joint replacements

4% Obstetrics, including pregnant women, new mothers and young children

2% Trauma including road accidents

Source: Bloodhound Study Monash Institute of Health Services Research 2007

Source: Bloodhound Study Monash Institute of Health Services Research 2007. With permission from Australian Red Cross Blood Service.

a What does each person symbol stand for?

b Explain how this graph helps you understand how blood donations are used.

c Is there any information in this graph that surprises you? Explain your answer.

d Does this graph make you more likely to donate blood?

e If 27 000 litres of blood are collected in one week, how much of it goes to help patients with cancer and blood diseases?

INVESTIGATION

GRAPHS IN OTHER SUBJECTS

Graphs are used to record and present data across all subject areas. For each of the subjects you study other than Mathematics:

* find examples of graphs used in the subject (find as many different types as possible)
* name the types of graphs used.

Which is the most common type of graph used across different subjects? Suggest a reason why this type of graph is used frequently.

Write a short paragraph on the uses of graphs in your areas of study. Explain the similarities and differences. Suggest reasons why different subjects use different types of graphs.

Present your findings as a PowerPoint presentation or wall chart.

5.04 Choosing the best graph

We use different graphs for different types of data. In this exercise, you will learn some hints to help you choose the best graph for each type of data.

Exercise 5.04 Choosing the best graph

1 Categorical data is often represented by a **column graph**. Draw a column graph for the following data showing the make of car owned by the families of a group of Year 11 students.

Make of car	Frequency
Toyota	12
Hyundai	10
Holden	7
Mazda	5
Ford	8
Other	3

2 The media often use a **picture graph** because it is an attractive graph. They are best used when the data is not too spread out and precise accuracy isn't required.

The Cairns Tourism Board wants to increase its number of tourists in winter. It published the number of rainy days in June for a number of cities. Graph the data as a picture graph, using the symbol 🌧 to represent 4 rainy days.

City	Rainy days in June
Cairns	10
Melbourne	18
Adelaide	24
Sydney	15
Hobart	17
Perth	16
Brisbane	15

3 A **pie graph** is used to show each category as a fraction of the whole. This pie graph shows the causes of children's deaths one year in Australia.

a What percentage of the deaths were caused by drowning?

b What was the most common cause of death?

c Drowning, assault and traffic incidents are considered preventable causes of death. What percentage of deaths are from something that could have been prevented?

d If there were 1769 child deaths in total, how many were preventable?

Causes of death in children

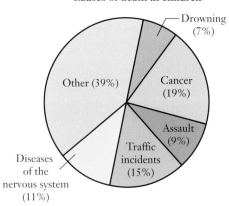

Drowning (7%)

Cancer (19%)

Other (39%)

Assault (9%)

Traffic incidents (15%)

Diseases of the nervous system (11%)

4 A survey of 1080 workers in central Melbourne were asked how they travelled to work.

Method of transport	Frequency
Train	220
Tram	280
Bus	130
Car	230
Walk	80
Cycle	140

a Copy and complete the following table which calculates the angle for each sector in a pie graph. Round each answer to the nearest degree.

Method of transport	Angle for sector graph
Train	$\dfrac{220}{1080} \times 360 =$
Tram	$\dfrac{280}{1080} \times 360 =$
Bus	$\dfrac{130}{1080} \times 360 =$
Car	$\dfrac{230}{1080} \times 360 =$
Walk	$\dfrac{80}{1080} \times 360 =$
Cycle	$\dfrac{140}{1080} \times 360 =$

b Draw a pie graph for this data.

5 This **line graph** charts Kate's height over her first 15 years. At birth, Kate was 48 cm long.

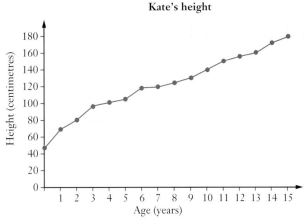

Kate's height

a What was Kate's height:

   i on her first birthday?

   ii at age 10?

b At what age did Kate reach:

   i 1 metre?

   ii 150 cm?

c Between which 2 birthdays did Kate grow the most?

d How long did it take Kate to double her height from birth?

e What do you think the graph would look like from age 15 to 20 years?

6 This table shows the temperatures over 24 hours in Alice Springs in November.

Time	Noon	3 p.m.	6 p.m.	9 p.m.	Midnight	3 a.m.	6 a.m.	9 a.m.	Noon
Temperature	32°C	35°C	33°C	29°C	26°C	24°C	23°C	29°C	34°C

Draw a line graph for this data.

7 This data shows the monthly number of burglaries over 3 years in an inner-city region.

54, 41, 55, 49, 37, 38, 37, 48, 51, 44, 52, 44, 58, 70, 60, 46, 63, 54
45, 43, 46, 55, 55, 67, 49, 66, 90, 45, 66, 62, 51, 51, 53, 53, 38, 52

Copy and complete the **stem-and-leaf plot** for this data.

Stem	Leaf
3	
4	
5	
6	
7	
8	
9	

Remember: a stem-and-leaf plot has the tens digit as the stem and the units digit as the leaves.

8 The water level of a canal was measured over 44 days.

Draw a **frequency histogram** for this data.

Water level (cm)	Frequency
152	5
153	6
154	8
155	12
156	4
157	8
Total	44

9 We use a **frequency histogram** or a **frequency polygon** to see the shape of the data. A group of students was given a maths test and the test marks out of 80 were grouped into class intervals, then graphed on this frequency histogram and polygon.

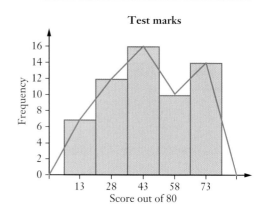

a Copy and complete the frequency table using the graph.

Test marks	Frequency
6–20	
21–35	
36–50	
51–65	
66–80	
Total	

b How many students sat the test?

c What was the most common class interval of test marks?

d How many students scored less than 36 in this test?

5.05 Graphs and spreadsheets

Statistical
graphs

In this exercise you will use a spreadsheet to produce some graphs. Ask your teacher to download the 'Statistical graphs' spreadsheet with the data for this exercise from NelsonNet. You should save this spreadsheet to your own computer.

Exercise 5.05 Graphs and spreadsheets

1 **Column graph**

To draw a column graph that shows the numbers of people immigrating to Australia by their region of origin in 1988–9:

- select all of the cells from A3 to B8
- select the **Insert** tab at the top
- under **Charts**, select **Column** and then the first of the 2D column types
- click on the chart title, highlight the text '1988–9' and change it to 'Immigration by region of origin'
- ensure the chart is selected (**Chart Tools** appears at the top of the sheet), select the **Layout** tab, then axis titles, go to **Primary Vertical Axis Title** and select **Rotated Title**. This will insert vertical script on the vertical axis. Highlight this script and change it to 'thousands'.
- with the chart selected, choose the **Design** tab and pick **Move Chart**. In the dialogue box, select the Object in option and immigration as the sheet and then OK. Move the whole chart to the top left of this sheet (click and hold the mouse on the chart border, move to the desired location and unclick).

2 **Clustered column graph**

Several years of information can be provided in the one graph by using clustered columns. For example, data for both years 1998–9 and 2008–9 can be shown.

- Return to the data sheet.
- Select all cells from B3 to C8.
- Select the **Insert** tab, and under **Charts** choose **Column** and select the first one of the 2D column types.
- Go to the **Layout** tab, click **Chart Title** and select **Above Chart**. Change the title text to the same as for the previous chart.
- Give the vertical axis the label 'thousands' as for the previous chart.
- Move the chart to the immigration sheet as you did for the last one and place it just below the previous column chart.

3 Find the latest data for immigration by region of origin and add a third column to the data provided. Draw a clustered column graph for all 3 years. Place the graph on the immigration sheet.

4 **Pie graphs**

The same data will be used to draw sector charts of the immigration statistics to show the contribution from the various regions.

- Return to the data sheet and select cells A2 to B8.
- Select the **Insert** tab, and under **Charts**, choose **Pie** and select the first 2D pie type.
- Check out the different possible **Chart Layouts** (top of screen). Try them out and choose the chart that you think shows the most information in the way that looks best.
- Move the chart to the immigration sheet as you have done previously and place it just to the right of the first simple column chart.

    Now create another sector chart for the 2008–9 data. The steps are much the same as before, except the initial selections of cells is a little more complicated.
- Return to the data sheet. Select cells A3 to A8. Then press and hold the Ctrl button while you select C3 to C8 with the mouse.
- Insert the pie chart as before.
- This time the title needs to be changed. Click the cursor in front of 2008–9 in the title and add 'Immigration to Australia by Region of Origin' in front of it.
- When you are happy with its format, place it below the first sector chart on the immigration sheet.

5 Line graphs

Use the data for temperatures during a spring day in Launceston in cells A24 to J25 of the data sheet.

- Select cells A24 to J25
- Choose **Insert**, **Line** and select the first of the 2D chart types.
- Go to the **Layout** tab, go to **Chart Title** and select **Above Chart**. Replace the words Chart Title with 'Temperatures in Launceston'.
- Place the graph in the 'lines' worksheet.

6 Use the vehicle and home theft figures in cells A10 to F13 of the data sheet. Both will be plotted on the same graph. Draw a line graph for this data. Place the graph in the 'lines' worksheet.

7 Use the data on Bazza's gym from the data sheet (A17 to M20). Draw a line graph for this data. Place the graph in the 'lines' worksheet.

5.06 Line graphs

Line graphs are usually used to graph data that changes over time. They can also be used to compare 2 or more sets of data.

Exercise 5.06 Line graphs

1 Simon was born weighing 4.3 kg. This line graph shows his weight over his first 12 months.

a What did Simon weigh at:

i 2 months?

ii 7 months?

b When did Simon weigh:

i 10 kg?

ii 12 kg?

c Between which 2 months did Simon gain the most weight?

d Approximately how long did it take Simon to double his birth weight?

e Suggest what this graph might look like over the next 12 months. Give reasons for your suggestion.

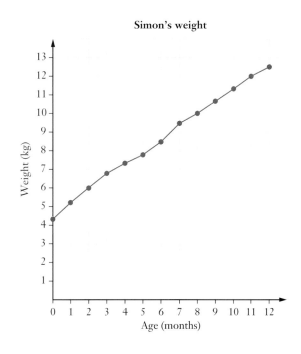

Simon's weight

2 These graphs show the percentages of deaths from cancer and heart disease at different ages.

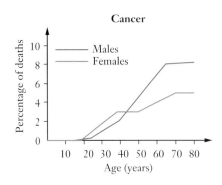

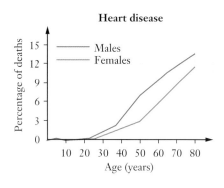

a For males aged 70, what percentage of deaths are due to cancer?

b Between what 2 ages do 3% of women die of cancer?

c For males aged 50, what percentage of deaths are caused by heart disease?

d Generally, according to both graphs, what happens as people get older?

e Is the statement 'More men die of heart disease than women' true or false? Suggest a reason why.

f At what age does the rate of cancer-related death in men increase?

g Between what ages do more women die of cancer than men? Discuss why.

3 The 8 graphs below illustrate the average maximum (orange) and minimum (blue) monthly temperatures of the Australian state capitals (Adelaide, Brisbane, Canberra, Darwin, Hobart, Melbourne, Perth and Sydney) over a 12-month period, but they are presented in jumbled order.

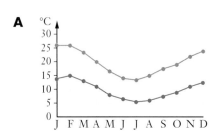

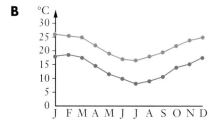

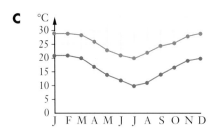

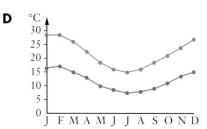

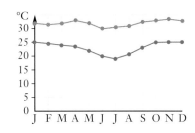

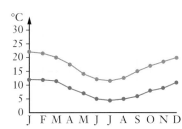

a From your knowledge of geography and climate, decide which graph belongs to which city.

b For graph G, list the maximum temperature of each month.

c Graph D has two months with exactly the same minimum temperature. Which months are they?

d Find which graphs have one month with a minimum of 10°C. For each graph, state which month it is.

e Which graph has all the maximum temperatures approximately the same?

f Which graph has the greatest difference between the highest maximum temperature and the lowest minimum temperature for one month? What is this difference?

g Find the highest maximum temperature and state which graph shows it and in which month it occurs.

h Find the lowest minimum temperature and state which graph shows it and in which month it occurs.

4 This line graph shows the midday temperature for the last 20 days of April.

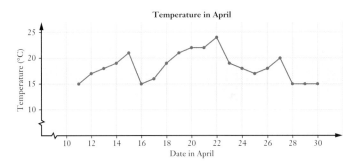

a On what date was the highest temperature, and what was that temperature?

b What was the lowest temperature?

c What was the range (the difference between the highest temperature and the lowest temperature)?

d What is the scale shown on the vertical axis?

e What was the temperature on 17 April?

f On what days was the temperature 22°C?

g Which day experienced the biggest drop in temperature?

h On how many days was it colder than the day before?

5 Sarah is in hospital recovering from having her tonsils removed.

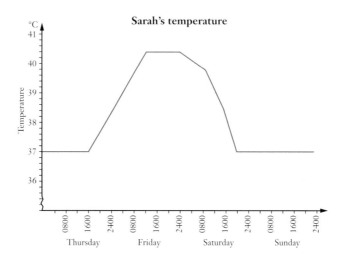

a What does each unit represent on:

   i the time scale?   **ii** the temperature scale?

b What was Sarah's temperature at 0800 Friday?

c When did Sarah's temperature first reach 38.3°C?

d What was Sarah's maximum temperature?

e How long was her temperature at this maximum value?

f Increases in temperature can indicate an infection. Sarah's doctor prescribed some antibiotics. When did her temperature return to the normal 37°C?

6 Angelina was admitted to hospital at 10 p.m. with a high temperature. This table shows her temperature each hour for the next 12 hours.

Time	10 p.m.	11	12	1 a.m.	2	3	4	5	6	7	8	9	10 a.m.
Temperature (°C)	42	42	41	40	40	41	39	40	38	39	37	37	37

Draw a line graph showing this information.

7 Jacob and Evan have decided to both diet and exercise in order to lose weight. The table records their weight loss over a 10-week period.

a Copy and complete the table below.

	Week	Start	1	2	3	4	5	6	7	8	9	10
Jacob	Weight loss (kg)		4	7	5.5	12		5	6.5	8.5	3	
	Weight (kg)	183					152					125
Evan	Weight loss (kg)		2.5	4.5	1.1		6.4	3	4.6	5	5	
	Weight (kg)	137			125							99

b On one graph, draw two line graphs showing Jacob and Evan's weight loss over 10 weeks. Make each line graph a different colour.

c On one graph, draw two line graphs of Jacob and Evan's actual weight over 10 weeks. Make each line graph a different colour.

d By looking at both the sets of graphs you have drawn, who do you think was more successful at losing weight? Justify your choice.

e What is the total weight loss for
  i Jacob?   **ii** Evan?

f Calculate the total weight loss as a percentage of the starting weight for both Jacob and Evan. Give your answer correct to 1 decimal place.

5.07 Misleading graphs

Sometimes graphs are deliberately drawn to give a false impression, like the one shown. This can be done by:

- not having a scale on the vertical axis

- showing only part of the scale or an irregular scale

- not showing zero on the vertical scale or only showing a small part of the vertical axis

- using pictures or three-dimensional figures on the graph to exaggerate the differences

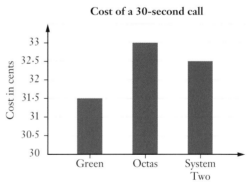

The graph shows the cost of a 30-second call on three different mobile phone plans. If you view the graph quickly, it appears that the calls are much cheaper on the Green plan than on the other two plans, but a closer look will show that there really isn't a big difference in cost between the three plans. While the information on the graph is correct, it gives a **misleading** impression.

Only showing a small part of the vertical axis on a graph is the most common way to create a misleading impression. As you work through the questions in the next exercise, you will learn about other ways to create misleading impressions.

ISBN 9780170412650

5. Show me the graph   117

Exercise 5.07 Misleading graphs

1 Use the misleading graph on the previous page to answer the following questions.

   a How much does a 30-second mobile phone call cost on each of the 3 plans?

   b How much cheaper is a call on the Green plan than on the System Two plan?

   c This graph shows the same information as the misleading graph.

   In which graph does the cost of a call on the Green plan look much cheaper than the cost on the other plans?

   d Write a sentence to describe how the scales on the vertical axes are different on the 2 graphs.

   e Which company do you think produced the misleading graph? Give a reason for your answer.

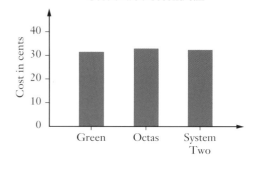

2 David drew a column graph to show the average cost of a litre of petrol in 6 capital cities.

   a In which city is the price of petrol the cheapest?

   b Which city has the highest petrol price?

   c Use the graph to estimate the price of petrol in Brisbane.

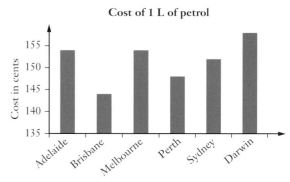

   d Approximately how many cents cheaper is the price of petrol in Brisbane than in Darwin?

   e David included this graph in an advertisement for motoring holidays in Brisbane. He claimed that the graph shows that petrol prices are much lower in Brisbane than they are in other capital cities. Explain what David did to the graph to make the price of petrol look a lot cheaper in Brisbane.

3 This table shows the price of a brand of coffee in 2010 to 2015, with the data graphed in 2 different ways on the next page.

Year	2010	2011	2012	2013	2014	2015
Price of coffee	$6.15	$6.60	$7.30	$8.05	$8.90	$9.65
Increase since 2010	$0	$0.45	$1.15	$1.90	$2.75	$3.50

Price of coffee since 2010

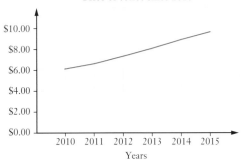

Increase in price of coffee since 2010

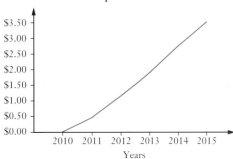

The 2 graphs illustrate how graphs can make **increases** look small or big. Imagine that you want to draw a graph to show how the population in your area has increased. Describe how you could make the increase look small and how you could make it look big.

Shutterstock.com/Africa Studio

4 Scott thinks that oil companies are ripping people off. He decided to investigate the price of petrol and the cost of crude oil. Scott displayed his information on a graph.

PS

Petrol and crude oil prices

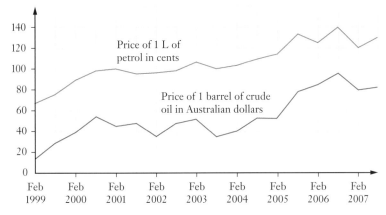

a What was the price of 1 L of petrol in February 2007?

b How much did a barrel of crude oil cost in February 2007?

ISBN 9780170412650

c Profit = selling price – cost price. The graph shows the barrel cost at 56 and the petrol cost at 104 in February 2003, giving the impression that the profit is 104 – 56 = 48. Explain why this impression is completely false.

d What relationship does the graph show between the price of petrol and the price of crude oil?

5 This graph compares the average weekly wages in Malvolia ($700) and Australia ($1400).

a What misleading impression does this graph give?

b Explain two things that are wrong about this graph.

c What should be drawn on the graph instead of pictures?

d Redraw this graph correctly.

Average weekly wages

Malvolia   Australia

6 This question uses data from the spreadsheet from Exercise 5.05. It uses the results of a survey of households about their favourite commercial television network. The data are in cells A37 to B41.

a Select cells A38 to B41 and make a basic column chart from the data.

b Select **Layout**, then **Axes, Primary Vertical Axis, More Primary Vertical Axis, Options**.

c In the dialogue box for **Minimum**, select **Fixed** and enter 3700 next to it. Close the dialogue box.

d Give the graph the obvious title it deserves 'Nine blitzes the other networks' and put it in the misleading sheet.

e Do the same again but in the vertical axis dialogue box make the minimum zero. Close the box.

f This is the same data but the impression is very different. Give this graph a title that explains the situation seen.

g Move it to the misleading sheet directly beneath the first one.

KEYWORD ACTIVITY

In your own words, summarise the important features of each type of graph.

1 Column graph

2 Climate graph

3 Clustered column graph

4 Conversion graph

5 Histogram

6 Line graph

7 Picture graph

8 Pie graph

9 Stem-and-leaf plot

NELSON QMATHS 11. Essential Mathematics

ISBN 9780170412650

SOLUTION ᵀᴼ ᵀᴴᴱ CHAPTER PROBLEM

During a drought, the water authority produced a graph to show the decreasing supply of water in a dam due to the water usage of local residents. Is the decrease in water supply as bad as the graph shows? If the graph gives a false impression, identify what has been done to create this impression.

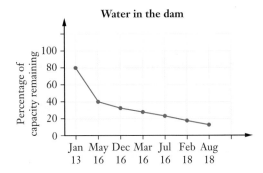

Solution

STAGE 1: WHAT IS THE PROBLEM? WHAT DO WE KNOW?

We need to decide if the graph is misleading and, if so, why?

WHAT?

STAGE 2: SOLVE THE PROBLEM

This graph is misleading because the scale on the horizontal axis isn't consistent. The horizontal scale jumps from 2013 to 2016 and then various months from 2016 and 2018.

SOLVE

The big drop in dam levels between January 2013 and May 2016, compared with the small drop between February and August in 2018, gives the impression that water was being wasted from 2013 to 2016.

STAGE 3: CHECK THE SOLUTION

We have answered both parts of the question.

CHECK

STAGE 4: PRESENT THE SOLUTION

The graph below shows the same information in a non-misleading way.

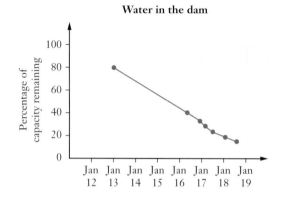

5. CHAPTER REVIEW

Show me the graph

Use the graphs in Exercise 5.01 on pages 92–99 to answer questions **1** to **5**.

1 Use the picture graph in question **1**.
   a How many cars passed the school gate between 8 a.m. and 10 a.m.?
   b During which time period did 5 cars pass the school gate?

2 Use the column graph in question **3**.
   a Estimate the population of Brisbane from the graph.
   b Estimate the difference between the populations of Brisbane and Melbourne.

3 Use the line graph in question **7**.
   a What was the home loan interest rate in June 1991?
   b By how much did the home loan interest rate rise from 1994 to 1995?

4 Use the conversion graph in question **9**.
   a Convert $60 to euros.
   b Convert 60 euros to Australian dollars.

5 Use the pie graph in question **14** of Exercise 5.01.
   a Which sport has the smallest sector?
   b Football was the favourite sport of 17% of those surveyed. 3500 people were surveyed. How many people preferred to watch football on TV?

6 The government gathered information about people's internet access by undertaking a survey. The following table shows the results of that survey.

	High-speed internet	No high-speed internet	
City	105	38	i
Country	89	42	ii
	iii	iv	v

   a Copy the table and complete the missing values **i** to **v**.
   b How many country people were surveyed?
   c What percentage of city people have high-speed internet?
   d What percentage of the whole sample do not have high-speed internet?

Use the graphs in Exercise 5.03 on pages 103–107 to answer questions **7** to **9**.

7 Use the telephone bill graph in question **1**.

   a How much was the bill in August?

   b What is the difference between the highest telephone bill and the lowest telephone bill over this 13 months?

8 Use the gas bill in question **3**.

   a What is the lowest usage per day for gas shown on this graph?

   b When did this occur?

   c Suggest why the gas usage would be lowest at this time.

9 Use the table of interest rates in question **4**.

   a What is the highest interest rate offered?

   b Annabel invests $50 000 for 6 months. How much interest does she earn?

10 Copy and complete this paragraph:

A column graph is often used for _____ data. When we are showing parts of a whole, we use a _____ graph. A line graph is used to show _____ over _____. An advantage of a stem-and-leaf plot is _____. To see the shape of a data set, we use a _____.

11 This table gives the daily minimum temperature over a fortnight in Innisfail.

Day	1	2	3	4	5	6	7	8	9	10	11	12	13	14
Temp (°C)	20	19	21	21	20	19	18	20	17	16	19	17	16	14

Draw a line graph showing this information.

12 Redraw the misleading graph in Exercise 5.07, question **2** on page 118 so that it isn't misleading.

Section A Multiple-choice questions

For each question select the correct answer **A**, **B**, **C** or **D**.

1 Andrew's bank account balance was –$44, meaning he owed the bank $44.
He deposited $120 into his account. What is his balance after the deposit?

A $164   **B** $76   **C** $74   **D** $64

Exercise 1.03

2 The cost of sending a parcel weighing 5.5 kg to Malaysia is $88. What is the postage rate?

A $0.07/kg   **B** $1.60/kg   **C** $16/kg   **D** $484/kg

Exercise 4.01

3 In one season, a soccer team scored 20 goals. Kirsten scored 4 of them. What is this as a percentage of the team score?

A   4%   **B**   5%   **C**   20%   **D**   40%

Exercise 2.02

4 What type of graph is best for displaying the mass of a baby changing over time?

A   line   **B**   column   **C**   sector   **D**   picture

Exercise 5.04

5 A group of people were surveyed about whether they exercised regularly.
This incomplete table shows the results of the survey.

Exercise 5.02

Age group	Exercise	No exercise	Total
Under 40	60		110
40 and over		25	
Total	125	75	

How many people were surveyed altogether?

A   90   **B**   110   **C**   125   **D**   200

6 Which decimal below can be rounded to 6.74?

A   6.7349   **B**   6.732   **C**   6.7452   **D**   6.744

Exercise 1.07

7 A department store has a mark-up of 200% on clothing. The store buys a vest for $12. What will be its selling price after the mark-up?

A $24   **B** $36   **C** $212   **D** $224

Exercise 2.03

8 Tuna cat food is sold in 4 different packs. Which pack is the cheapest price per 100 g?

A   24 × 250 g cans for $39.99   **B**   12 × 300 g cans for $24.99

C   9 × 400 g cans for $25.99   **D**   6 × 1 kg cans for $38.50

Exercise 4.03

Section B   Short-answer questions

Exercise 4.04

1 A tap drips water at a rate of 18 mL/h. How much water is wasted from these drips over a week?

Exercise 2.01

2 Find:

   a 2% of 250 kg

   b 75% of $150 000

   c 80% of 4 hours (in minutes)

   d 18% of 36 L (in mL)

Exercise 5.01

3 This graph shows the number of motor vehicle thefts per year in different areas.

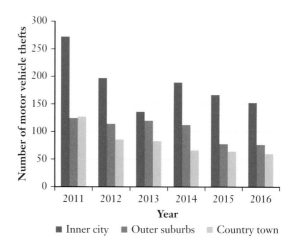

   a Which year had the highest number of vehicle thefts in the outer suburbs?

   b In which area have motor vehicle thefts declined the most?

   c In which year was the difference between the number of thefts in the inner city and the country towns the highest? Estimate this difference.

Exercise 1.04

4 Evaluate each expression.

   a $2 + (13 - 6) \times 11$

   b $\dfrac{8 - 3}{100 \div 4}$

   c $14 - 5 \times (-3)$

   d $\dfrac{4^2}{16 \div (-2)}$

Exercise 2.02

5 What percentage is:

   a 69 out of 75?

   b 9 mm out of 300 mm?

   c 300 g of 2 kg?

   d 6 h of 1 day?

ISBN 9780170412650

6 Round each quantity to the nearest whole number.

   a   85.4 m   **b**   16.9 kg   **c** $9.80

Exercise 1.07

7 a Copy and complete this table from Section A, Question 5.

Age group	Exercise	No exercise	Total
Under 40	60		110
40 and over		25	
Total	125	75	

Exercise 5.02

   b   What percentage of the people surveyed do not exercise regularly?

   c   Of those who exercise regularly, what percentage were aged 40 and over?

8 Judy's car uses diesel fuel and has a fuel consumption of 7.4 L/100 km.

   a   How far can Judy travel on 50 litres of diesel fuel? Answer to the nearest 5 km.

   b   Diesel costs 129.9c/L. How much does it cost Judy to put 50 L of diesel fuel into her car?

Exercise 4.05

9 At the end-of-financial-year sale, a car yard offers an 18% discount on all cars. Calculate the price of a car marked at $29 990, correct to the nearest dollar.

Exercise 2.04

10 Draw a line graph for this data about the temperature over 1 day in winter.

Time	9 a.m.	10 a.m.	11 a.m.	12 noon	1 p.m.	2 p.m.	3 p.m.	4 p.m.
Temperature	8°C	11°C	13°C	16°C	18°C	17°C	16°C	13°C

Exercise 5.06

11 Terry and Andrea purchased a block of land for $133 000. Six years later they sell it for $164 000. Calculate their percentage profit correct to 2 decimal places.

Exercise 2.05

12 A large vat is being filled at the rate of 35 mL/min. Convert this to a rate in L/day.

Exercise 4.02

6.

COLOURFUL RATIOS

Chapter problem

Lara needs 24 mL of purple paint for a flower scene she is painting. To make the shade of purple she wants, she will mix some cyan and magenta paint in the ratio of 1 : 2. How much of each colour should she use?

WHAT WILL WE DO IN THIS CHAPTER?

- Understand the meaning of a ratio
- Write and simplify ratios
- Solve problems involving ratios and scales
- Divide an amount in a given ratio
- Use scale ratios in scale drawings

HOW ARE WE EVER GOING TO USE THIS?

- Mixing paint, hair colour and cake ingredients
- Mixing fertiliser, bleach, pesticide and other concentrates
- Interpreting scale drawings, models and plans

Ratios

Ratio word blanks

Ratio match-up puzzle

Ratios code puzzle

6.01 Ratios

A **ratio** shows the relative sizes of 2 or more quantities of the same type. Before we can write quantities as a ratio they must be in the **same units**.

Write the ratio for:

a   15 cm to 37 cm

b   59 seconds to 2 minutes.

Solution

a   Both quantities have the same units, cm.   Ratio is 15 : 37.

b   These quantities have different units, so convert 2 minutes to seconds.   2 minutes = 2 × 60 = 120 seconds.

Now, both quantities are in seconds.   Ratio is 59 : 120.

To **simplify** a ratio, divide each term in the ratio by the same amount.

Simplify each ratio.

a   10 : 6

b   21 : 7

c   10 : 15 : 20

Solution

a   Both 10 and 6 can be divided by 2.

We can also use the fraction key on our calculator: enter 10 a^b/c 6 $=$ and the calculator will display $\frac{5}{3}$, which we write as 5 : 3.

$$10 : 6 = 10 \div 2 : 6 \div 2$$
$$= 5 : 3$$

b   Divide both 21 and 7 by 7.

For 21 a^b/c 7 $=$, the calculator will display 3, which we can write as 3 : 1.

$$21 : 7 = 21 \div 7 : 7 \div 7$$
$$= 3 : 1$$

c   All 3 terms can be divided by 5.

When there are more than 2 terms in the ratio, we can't use the calculator to simplify.

$$10 : 15 : 20 = 10 \div 5 : 15 \div 5 : 20 \div 5$$
$$= 2 : 3 : 4$$

We should always check our answer to see if it can be simplified further.

Exercise 6.01 Ratios

1 Write the ratio for:

   a 13 g to 25 g   **b** $17 to $100   **c** 5 litres to 12 litres

   d 150 km to 1 km   **e** 31 seconds to 51 seconds to 101 seconds.

Example 1

2 Marianne surveyed the eye colour of the 25 members of her netball club and found that 12 had brown eyes, 8 had blue eyes, 4 had green eyes and one had hazel eyes.
Write the ratio of players with:

   a hazel eyes to green eyes   **b** brown eyes to hazel eyes

   c blue eyes to brown eyes   **d** brown eyes : green eyes

   e blue eyes : brown eyes : hazel eyes   **f** green eyes to blue eyes to brown eyes.

3 Kris buys avocados at the market for $2 each and sells them in his grocery store for $3 each.

   a How much profit does Kris make on each avocado?

   b What is the ratio of:

      i selling price to cost price?   **ii** selling price to profit?

      iii cost price to profit?   **iv** profit to selling price?

4 Of the 75 people who attended a bush dance, 27 were men, 29 were women and the remainder were children. Find the following ratios:

   a men to women   **b** children to men   **c** women to the total

   d women to children   **e** men to women to children.

5 Write the ratio for:

   a 3 cents to $2.50   **b** 23 mm to 3 cm

   c 37 minutes to 2 hours   **d** 10 years to 36 months

   e 14 days to 5 weeks   **f** 0.3 km to 173 metres.

6 The diagram shows a cluster of yellow, green, purple, grey, red and black balls. Write the colours that are in the ratio:

   a 6 : 1   **b** 4 : 3   **c** 2 : 5

   d 2 : 21   **e** 2 : 1 : 6   **f** 3 : 4 : 5

> Be sure to write the colours in the same order as the ratio is written.

7 Simplify each ratio.

Example 2

   a 8 : 6   **b** 9 : 12   **c** 15 : 10   **d** 20 : 10

   e 21 : 14   **f** 27 : 9   **g** 18 : 12   **h** 150 : 25

   i 120 : 24   **j** 56 : 49   **k** 28 : 52   **l** 200 : 50

   m 8 : 12 : 4   **n** 6 : 3 : 18   **o** 15 : 25 : 50   **p** 90 : 60 : 120

8 Simplify each ratio using your calculator.

a $4.5 : 9$ **b** $1.8 : 2.7$ **c** $2\frac{1}{2} : 7\frac{1}{2}$ **d** $14 : 3\frac{1}{2}$

e $\frac{1}{2} : \frac{3}{4}$ **f** $2.9 : 5.8$ **g** $3.5 : 2.1$ **h** $3.6 : 1.2$

9 Karen uses a soil conditioning solution on her vegetable garden. The ingredients are listed on this label shown.

Express in simplest form the ratio of:

a seaweed extract to water

b organic acid to water

c seaweed extract to organic acid

d seaweed extract to total solution

e seaweed extract to organic acid to water.

FERTILISER

Ingredients

Seaweed Extract 135 mL
Organic Acid   45 mL
Water   900 mL

10 The table shows the average annual yield in kilograms from various fruit and nut trees.

Apple	144		Nectarine	63
Apricot	36		Fig	9
Pear	54		Walnut	27

What is the yield ratio of:

a apple to pear?   **b** nectarines to walnuts?

c apricot to apple?   **d** pear to fig?

e apples to nectarines to figs?

11 Steven runs a small gourmet business producing specialty fruits, vegetables and cereal products. He also has poultry producing a variety of eggs. The drawing shows the area of land he has dedicated to each activity. Find the ratio of the following production areas.

a fruit to vegetables

b egg to total area

c vegetables to total crop area (the area excluding poultry)

d cereal crop to the total crop area

e vegetable to fruit to cereal crop area.

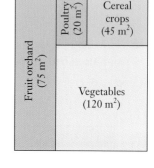

Fruit orchard (75 m²)

Poultry (20 m²)

Cereal crops (45 m²)

Vegetables (120 m²)

12 Robyn is cleaning the floor with a cleaning liquid she has mixed from a detergent concentrate and water. The instructions say to mix concentrate to water in the ratio 3 : 16 for light cleaning or 5 : 16 for heavy-duty cleaning. Robyn has mixed 22.5 mL of concentrate with 72 mL of water for this job.
Is she doing light or heavy-duty cleaning?

6.02 Mixing paint

Lara uses ratios to make the colours she needs for her paintings. She knows that the order of a ratio is very important. If she gets it wrong, the resulting colour is wrong and she wastes paint.

Lara mixed yellow and blue paint. When she used the ratio of 5 : 1, she mixed 5 parts yellow with 1 part blue. When she used the ratio of 1 : 5, she mixed 1 part yellow and 5 parts blue. The images here show that the resulting 2 colours were quite different.

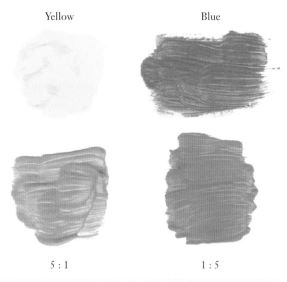

Yellow   Blue

5 : 1   1 : 5

Ratio problems

EXAMPLE 3

Lara mixed 12 mL of white paint with 8 mL of red paint to make dark pink paint.

a What is the ratio of red to white paint in the mixture?

b What fraction of the mixture was red paint?

Solution

a The order is important. For red : white we need to write the quantity of red first.

Red : white = 8 : 12

b 12 + 8 = 20 mL of total paint and 8 mL came from red paint. That's 8 mL out of 20 mL.

Fraction of the paint that was red paint $= \dfrac{8}{20}$.

EXAMPLE 4

Zack is painting his lounge room. He wanted the colour to be white with a hint of brown. He mixed 10 mL of brown with 8 L of white paint. What ratio of brown to white should he write on the lid of the paint tin so that he can make the same colour in the future?

Solution

Change 8 L into mL so the units are the same.

8 L = 8000 mL

Write the brown amount first.

Brown : white = 10 mL : 8000 mL

= 10 : 8000

Simplify the ratio.

= 1 : 800

Exercise 6.02 Mixing paint

Example
3

1 Lara mixed 5 mL of red with 9 mL of green paint to make brown to create a painting of some trees.

    a What is the ratio of green to red paint in the brown mixture?

    b What fraction of the brown paint is red?

2 Lara mixed 6 mL of green with 2 mL of blue paint to make a blue-green paint for the leaves.

    a What is the ratio of green : blue paint in the blue-green mixture?

    b Simplify your answer to part **a**.

    c What fraction of the blue-green mixture came from the green paint?

3 Lara made 2 containers of orange paint by mixing red and yellow. She mixed 20 mL of red with 40 mL yellow in one mixture and 8 mL of yellow with 4 mL of red in the other. She is confused. Both colours are the same! Why?

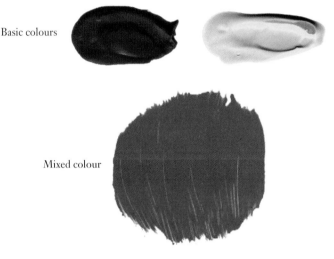

Basic colours

Mixed colour

Example
4

4 Zack mixed 5 mL of brown and 2 L of white paint to paint the woodwork in his lounge room.

    a What is the ratio of brown to white paint in mL?

    b Write your answer to part **a** in its simplest form.

iStock.com/Highwaystarz-Photography

5 Lara needs to make a particular shade of blue-green she wanted by mixing blue and green in the ratio of 4 : 1, but when she mixed the paint she got the wrong colour (see images). What could she have done wrong?

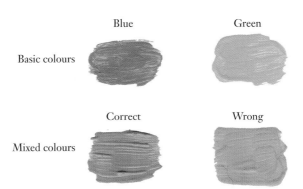

Blue

Green

Basic colours

Correct

Wrong

Mixed colours

6 Lara makes colours lighter by mixing them with white; the more white, the lighter the colour. Here are 4 shades of blue Lara made by mixing blue and white in the ratios of 3 : 1, 3 : 2, 1 : 1 and 1 : 2 but she forgot to label the diagrams. Match the ratios to the colour samples.

a   **b**   **c**   **d**

7 Lara makes this shade of purple by mixing red and blue paint in the ratio of 3 : 1. How much red paint does she need to add to 24 mL of blue to make this shade of purple?

8 Lara makes different shades of pink paint by mixing red and white, as shown in the table.

Calculate the amount of white paint Lara should add to 5 mL of red to produce each shade of pink.

	Colour sample	Ratio red : white
a		1 : 1
b		1 : 2
c		1 : 3
d		1 : 4

9 Compared to adults, children have bigger heads in relation to their bodies. This table shows values for males of different ages.

Age	head length : height	Typical height in cm
1 year	1 : 4	60
3 years	1 : 5	81
5 years	1 : 6	105
Adult	1 : 8	180

a What fraction of a 1-year-old boy's height is the length of his head?

b Lara is painting a picture of 5-year-old Tyson. In the picture, Tyson is 12 cm tall. How long should she make his head in the picture?

c What is the ratio of an adult man's height to the height of a 1-year-old boy?

d Lara is sketching a picture of a man with his 1-year-old son. In the picture, the man is 20 cm tall, so what length should Lara make:

  i the man's head? **ii** the son's height?

10 To make a fertiliser for his rose garden, Ahmed mixes manure and grass clippings in the ratio of 3 : 2.

a How many spades of grass does Ahmed need to mix with 15 spades of manure?

b When he uses 6 spades of grass, how much manure will he need to mix with it?

11 Nickel brass is a metal that contains copper, zinc and nickel mixed in the ratio 14 : 5 : 1. It is used to make many musical instruments as well as the English pound coin.

a A manufacturer used 70 g of copper to make a batch of nickel brass. How much zinc and nickel did he need?

b Vinh has 8 g of nickel. How much copper and zinc will he need to mix with it to make nickel brass?

c How much nickel brass can Vinh make?

Shutterstock.com/KenDrysdale

12 When Padmina makes scones for the school fete, she mixes self-raising flour and milk in the ratio 3 : 1.

a How much milk does she add to 15 cups of flour?

b If she uses 6 cups of milk, how many cups of flour are required?

13 Abi makes a spray to kill weeds by mixing a concentrated solution of poison with water.

Weed	Ratio of poison to water
Flat-leaf weeds	1 : 10
General perennial weeds	3 : 20
Woody weeds	1 : 50

a Abi needs to kill some blackberries, which are woody weeds. How many millilitres of water should she mix with 40 mL of poison to make the spray?

b How many litres of water is this?

c Abi is going to clear a small area infested with perennial weeds to make a new garden. How much poison does she need to add to 1 L of water to make a spray to kill the weeds? (Convert 1 L to mL first).

d Abi has flat-leaved weeds coming up in her lawn. How much poison should she mix with 250 mL of water to make the spray for these weeds?

e Abi has made a spray using 45 mL of poison mixed with 300 mL of water. For what type of weed will she use this spray?

PRACTICAL ACTIVITY: MAKING COLOURS

You need some red, yellow and blue paint (the **primary colours**), a surface for mixing paint and some paint brushes.

What you need to do

Find which primary colours need to be mixed to make each colour in the table below. Copy and complete the table with the correct colours and ratios. Purple has already been done for you.

Colour		Primary colour parts used	Ratio
		1 part red, 1 part blue	1 : 1
1			
2			
3			
4			

Ratios in other contexts

Dividing a quantity in a given ratio

Ratio problems

Calculating quantities in ratios

6.03 Dividing a quantity in a ratio

Sometimes we want to divide or share quantities into unequal amounts. For example, the owners of businesses often share the profits and expenses according to the size of their share in the business.

EXAMPLE 5

Lara mixes red, white and green paint in the ratio of 5 : 1 : 2 to make a red shade for painting house bricks. She needs to make 10 L of paint for the bricks. How much of each colour should she mix?

Solution

5 parts red, 1 part white and 2 parts green makes 8 parts.	$5 + 1 + 2 = 8$ parts
	8 parts make up 10 L
Find the size of one part by dividing 10 L by 8.	1 part = $10 \div 8$
	$= 1.25$ L
For each colour, multiply by the appropriate number of parts.	Amount of red = 5×1.25 (5 parts)
	$= 6.25$ L
	Amount of white = 1.25 L (1 part)
	Amount of green = 2×1.25 (2 parts)
	$= 2.5$ L

Check that your answer is correct by adding up the parts to see whether they make up the whole: $6.25 + 1.25 + 2.5 = 10$ L.

EXAMPLE 6

Janet and Darryl own a small market garden, which made a profit this year of $135 000. They have agreed to share the profit in the ratio 2 : 3. How much does each person receive?

Solution

Find the total number of parts. Find the size of one part.	Total number of parts = $2 + 3 = 5$. 1 part = $135\ 000 \div 5$
	$= \$27\ 000$
For each share, multiply by the appropriate number of parts.	Janet's share = $2 \times \$27\ 000$
	$= \$54\ 000$
	Darryl's share = $3 \times \$27\ 000$
	$= \$81\ 000$

Check that the parts add up to the whole: $\$54\ 000 + \$81\ 000 = \$135\ 000$.

Exercise 6.03 Dividing a quantity in a given ratio

1 Lara needs to mix blue and black in the ratio of 14 : 1 to make paint for a sky. She needs to make 60 mL. How much of each colour will she need?

2 Lara needs 24 mL of this shade of red to paint some waratahs in a bush scene. The colour is a 3 : 1 mix of red and yellow. How much of each colour will she need?

3 This shade of red will show the parts of the waratahs in the direct sunlight. It consists of red, yellow and white in the ratio of 4 : 1 : 1. Lara requires 12 mL of this colour. How much red, white and yellow paint should she mix together?

4 Lara needs 140 mL of a golden paint for the sand dunes in an outback scene. She will make the colour by mixing red, yellow and white in the ratio of 1 : 4 : 2. How much of each colour will she need?

5 At the end of each painting session, Lara cleans her equipment with commercial cleaner. She needs to dilute the concentrated cleaner with water in the ratio of 1 : 9, concentrate to water. How much concentrate will she need to make 4 L of diluted cleaner?

6 Lara sells her paintings in an art gallery. Her agent and she share the sale price in the ratio of 1 : 7. How much did Lara receive when one of her paintings sold for $4800?

7 Hamish needs to mix 180 kg of concrete. Concrete is made using cement, sand and gravel in the ratio 1 : 2 : 3. How much of each product will Hamish need?

8 Divide a prize of $2100 between Toby and Vinson in the ratio:

    a   2 : 3   **b**   5 : 1

9 At Jemma's school there are 741 students. The ratio of boys to girls is 7 : 6. How many boys are there in the school?

10 Asher is making a bracelet using silver beads, porcelain balls and crystal eyedrops in the ratio of 4 : 3 : 2. She uses 45 pieces to make the bracelet. How many of each type will she require?

11 A flat white coffee contains black espresso coffee and steamed milk in the ratio 1 : 2. How much milk is there in a 240 mL mug of flat white coffee?

12 Rose gold is used in jewellery and high-quality flutes because of its attractive appearance. It is an alloy of gold, copper and silver blended in the ratio 15 : 4 : 1.

 a How much silver is in a 40 g piece of rose gold jewellery?

 b A flute made of rose gold weighs 420 g. How much gold and copper does it contain?

13 Sam makes mocktails using 3 parts grape juice, 2 parts cranberry juice and 4 parts sparkling mineral water.

 a How many millilitres of grape juice will he need to make two 225 mL mocktails?

 b How much mineral water does Sam use to make 900 mL of the mocktail?

 c How many mL of cranberry juice does this mixture require?

6.04 Ratios of body parts

Artists use body part ratios to make sketches of people look realistic. The diagram shows some of the important ratios for sketching images of adults.

Men and women's bodies have a different shape. Women's necks and waists are thinner, while their hips and thighs are wider. This table shows some of the ratios that are different in adult males and females.

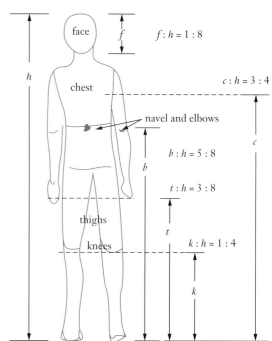

	Males	Females
Shoulder width : head length	7 : 3	2 : 1
Waist width : head length	15 : 7	1 : 1
Hips width : head length	2 : 1	3 : 2

EXAMPLE 7

Lara is sketching an adult male. Her sketch will be 20 cm tall. In her sketch, how big should she make each of the following features?

a The man's head length

b The height of his chest

c The width of his shoulders

Solution

a $f : h = 1 : 8$ from the diagram.

A person's head length is $\frac{1}{8}$ of his height.

Height in sketch = 20 cm.

$$\text{Head length} = \frac{1}{8} \times 20$$
$$= 2.5 \text{ cm}$$

b $c : h = 3 : 4$ from the diagram.

A person's chest height is $\frac{3}{4}$ of his height.

$$\text{Chest height} = \frac{3}{4} \times 20$$
$$= 15 \text{ cm}$$

c Shoulder width : head length = 7 : 3.

His shoulder width is $\frac{7}{3}$ times the length of his head (2.5 cm from part **a**).

$$\text{Shoulder width} = \frac{7}{3} \times 2.5$$
$$= 5\frac{5}{6} \text{ cm}$$

NELSON QMATHS 11. Essential Mathematics

ISBN 9780170412650

Exercise 6.04 Ratios of body parts

1 Lara is going to make a sketch of a man 12 cm tall. Calculate the size of each body part on the sketch.

a length of head **b** height of knees

c height of mid-thighs and fingertips **d** height of navel and elbows

e height of chest **f** width of shoulders

g width of hips

2 In the same sketch, Lara is going to draw a woman 10 cm tall. Calculate the size of each body part on the sketch.

a length of head **b** height of knees

c height of mid-thighs and fingertips **d** height of navel and elbows

e height of chest **f** width of shoulders

g width of hips

3 Use the answers to questions **1** and **2** to sketch a drawing of the man and woman.

WS

Scaled vs actual size

WS

Scales and scale diagrams

WS

Scale drawings

Scale drawings

6.05 Scale drawings

A **scale drawing** is a reduced or enlarged diagram of a real object, whose lengths are in the same ratio as the actual lengths of the object. The scale on a scale drawing can be given as a statement, for example, 1 cm represents 5 m, or a ratio, for example, 1 : 500.

The most common scale drawings are maps and house plans. By taking measurements on the scale drawing we can calculate the size of objects in real life using the scale given on the diagram.

To calculate a real length on a scale drawing:

- measure the scaled length on the scale drawing
- multiply by the scale factor
- convert your answer to the required units if necessary

EXAMPLE 8

Barry, a farmer, used a scale of 1 cm : 100 m when he drew this scale drawing of one of his paddocks.

a What are the actual dimensions of Barry's paddock?

b Express the scale as a ratio.

Solution

a Measure the scaled length.
Multiply by 100 m.

Scaled length = 4 cm
Actual length = 4 × 100 m
= 400 m

Measure the scaled width and multiply by 100 cm.

Scaled width = 2.5 cm
Actual width = 2.5 × 100 m
= 250 m

b Change 100 m to centimetres, then simplify.

Scale = 1 cm : 100 m
= 1 cm : 100 × 100 cm
= 1 cm : 10 000 cm
= 1 : 10 000

EXAMPLE 9

Olga, an architect, made a scale drawing of a house using a scale of 1 : 50.

a The actual length of the house is 20 m. What is the length of the house in the drawing?

b The width of the house in the drawing is 24 cm. What is the actual width of the house?

Solution

a Convert 20 m to cm first.

Actual length = 20 m
= 2000 cm

Divide by 50 for the scaled length.

Scaled length = 2000 cm ÷ 50
= 40 cm

b Multiply by 50 for the actual width.

Scaled width = 24 cm
Actual width = 24 × 50 cm
= 1200 cm
= 12 m

Exercise 6.05 Scale drawings

1 For each scale drawing, find the actual length shown by measurement and calculation, then express each scale as a ratio.

Example
8

a

Height

Scale:
1 cm represents 2 m

b

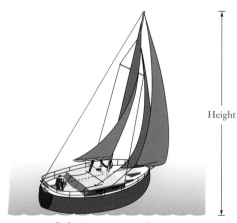

Height

Scale: 1 cm represents 5 m

c

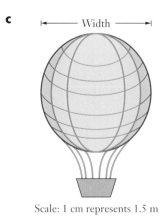

Scale: 1 cm represents 1.5 m

d

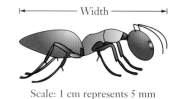

Scale: 1 cm represents 5 mm

e

Height

Scale: 1 cm represents 3 m

f

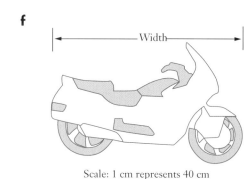

Scale: 1 cm represents 40 cm

2 A street map uses a scale such that 1 cm represents 200 m. Leanne walks from the Town Hall to the shopping centre, a distance that measures 5.5 cm on the map. How far does she walk?

3 Alan has a model of his favourite yacht in his study. The model has been made on a scale such that 2 cm represents 1 m. The model is 50 cm long. How long is the yacht?

 Example 9

4 Nagala used a scale of 1 : 50 when she constructed a scale drawing of her courtyard.

Find the length and width of Nagala's courtyard in metres.

5 Measure the length of each scaled image below, then use the scale ratio to calculate its actual length.

a Fish 1 : 3

Length

b House 1 : 300

Height

c Pen 1 : 4

Length

d Tennis racquet 1 : 16

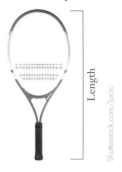

Length

6 Lena collects model cars made in the ratio 1 : 43. Her favourite model is 9.3 cm long. Calculate the actual length of the car to the nearest metre.

7 A map has a scale of 1 : 50 000. Dean has to travel a distance measured as 128 mm on the map. How far does Dean have to travel? Express your answer in kilometres.

KEYWORD ACTIVITY

1 What key words can you find in this chapter? Search through the chapter and list the keywords together with their meaning.

2 Use your own words to describe the process of dividing a quantity in a ratio.

SOLUTION TO THE CHAPTER PROBLEM

Problem

Lara needs 24 mL of purple paint for a flower scene she is painting.
To make the shade of purple she wants, she will mix some cyan and magenta paint in the ratio of 1 : 2. How much of each colour should she use?

Solution

WHAT?

STAGE 1: WHAT IS THE PROBLEM? WHAT DO WE KNOW?

To find paint quantities to mix.

We know ratio 1 : 2 (cyan : magenta)

Total amount required is 24 mL.

SOLVE

STAGE 2: SOLVE THE PROBLEM

There's a word clue:

Divide in a ratio → divide

Lara needs 1 part cyan to 2 parts magenta. That's 3 parts.

Each part will be 24 mL ÷ 3 = 8 mL

Lara will need 8 mL of cyan and 16 mL of magenta.

CHECK

STAGE 3: CHECK THE SOLUTION

8 mL + 16 mL = 24 mL and there's twice as much magenta.
✓ Correct!

PRESENT

STAGE 4: PRESENT THE SOLUTION

Lara needs 8 mL cyan and 16 mL magenta.

6. CHAPTER REVIEW

Colourful ratios

1 Lara mixed 24 mL of white paint with 8 mL of blue paint.

   a What ratio of blue to white paint did she use?

   b What percentage of the mixture was white paint?

Exercise
6.01

2 Simplify each ratio.

   a 10 : 25   **b** 8 : 4   **c** 50 : 20   **d** 300 : 500

Exercise
6.01

3 A recipe for making pizza dough mixes plain flour and water in the ratio 8 : 5. Jasmine uses 400 g of flour to make pizza. How much water does she add?

Exercise
6.02

4 Lara is mixing yellow and red paint in the ratio of 4 : 1. She is using 12 mL of yellow. How much red should she add?

Exercise
6.02

5 a Divide $80 between Manal and Eddie in the ratio of 4 : 1.

   b Divide 72 km in the ratio of 1 : 2.

Exercise
6.03

6 Lara needs 24 mL of a shade of blue that is obtained by mixing blue, green and white in the ratio of 3 : 2 : 1. How much of each colour should she use?

Exercise
6.03

7 The ratio of the length of a man's head to his total height is 1 : 8. Lara is sketching a man and makes his head 6 mm long. How long should she make the rest of his body?

Exercise
6.04

8 Measure the length of each drawing, and use the scale to find the actual length.

Exercise
6.05

   a

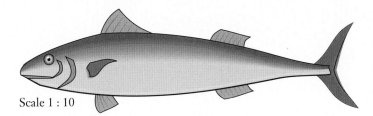

Scale 1 : 10

   b

Scale 1 : 4

PERCENTAGES

7.

APPLYING PERCENTAGES

Chapter problem

Aron buys some bedroom furniture on a store finance arrangement. The bedroom furniture costs $3200. The store finance will charge him 17% p.a. interest and expects him to pay it off monthly over 3 years. How much will Aron have to pay each month?

18.9% APR

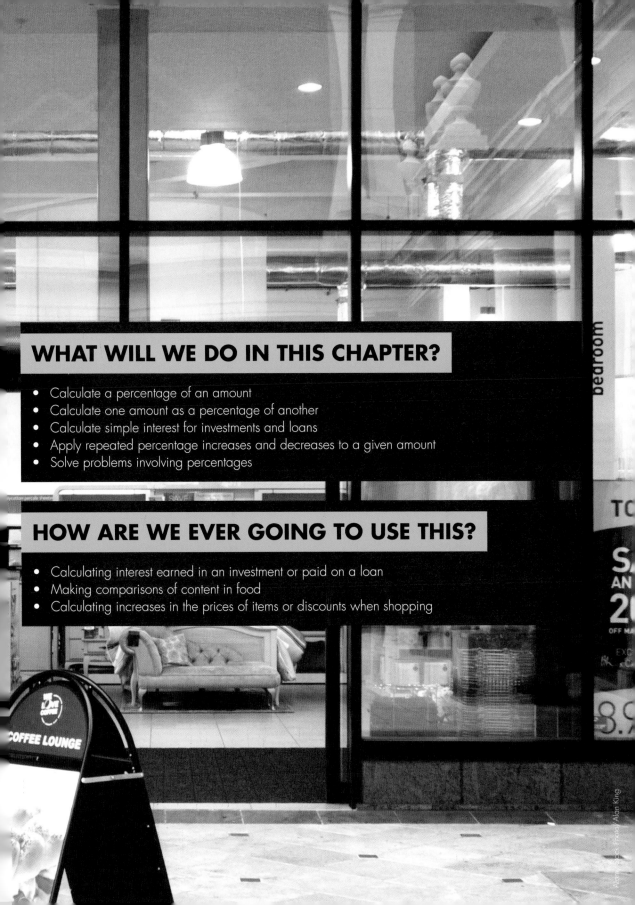

WHAT WILL WE DO IN THIS CHAPTER?

- Calculate a percentage of an amount
- Calculate one amount as a percentage of another
- Calculate simple interest for investments and loans
- Apply repeated percentage increases and decreases to a given amount
- Solve problems involving percentages

HOW ARE WE EVER GOING TO USE THIS?

- Calculating interest earned in an investment or paid on a loan
- Making comparisons of content in food
- Calculating increases in the prices of items or discounts when shopping

7.01 Finding a percentage of a quantity

Remember from Chapter 2:

$$\text{Percentage of a quantity} = \frac{\text{Percentage}}{100} \times \text{quantity} \qquad \text{OR}$$
$$= \text{Percentage} \div 100 \times \text{quantity}$$

EXAMPLE 1

Find:

a 74% of $63   **b** 40% of 2 km   ← 'of' means multiply.

Solution

a Write the percentage as a decimal (or fraction) and then multiply by the quantity.

74% of $63 = 0.74 × $63
= $46.62

b First convert 2 km to metres.

2 km = 2 × 1000 = 2000 m

Calculate the percentage.

40% of 2 km = 0.4 × 2000 m
= 800 m

Exercise 7.01 Finding a percentage of a quantity

1 Find:

a	12% of $125	**b**	50% of 88 000 people
c	7.5% of 14 kg	**d**	27% of 3000 letters
e	62.5% of 1200 m²	**f**	150% of 32 L
g	71% of 84 600	**h**	5% of 120 mins
i	0.4% of $5270	**j**	$6\frac{1}{4}$% of 960 people

2 Kelly went to Hong Kong for 28 days for a holiday. It rained on 25% of the days.

    a On how many days did it rain during Kelly's holiday?

    b How many days were fine?

3 In July, 93 700 new cars were sold in Australia. 27% of these were sold in Queensland. How many new cars were sold in Queensland in July?

4 Josephine earns $97.50 per week from her part-time job. She decides to save 30% of her income so she can go on holiday at the end of Year 12.

    a How much does she save each week? Round your answer to the nearest dollar.

    b Josephine is given a 2.5% wage increase. How much extra does she earn each week? Round your answer to the nearest cent.

5 Brandon pays $225 per week in rent for his share of a house. The landlord decides to increase the rent by 4%.

    a How much extra rent will Brandon have to pay?

    b What is the total amount of rent he will have to pay?

6 At Vanna's school, 85% of Year 11 students have a smartphone. There are 133 students in Year 11. How many have smartphones? Give your answer correct to the nearest whole number.

7 Will buys stationery for his business at *Office Perks*. He has to pay 10% GST on all items. Find the amount of GST he pays on items totalling $391.

8 New vehicle registration duty is a tax paid to the state government when you purchase a new car. In Queensland, it is 3% of the purchase price of the car. Simone buys a new car for $29 990. How much new vehicle registration duty will she pay?

9 An aircraft can seat 350 passengers. The plane flies from Brisbane to Perth with 82% of seats occupied. How many passengers were on the plane?

10 Australia has approximately 800 000 km of roads. Approximately 40.5% of Australian roads are sealed with bitumen or concrete. All other roads are gravel, sand or dirt.

    a How many kilometres of Australian roads are sealed?

    b Of the roads that are *not* sealed, approximately 66% are improved surfaces. How many kilometres of improved surface roads are there?

11 a Approximately 30% of Australians suffer from arachnophobia, the fear of spiders. In a school of 1200 students, how many students would you expect to suffer from arachnophobia?

    b Necrophobia is the fear of death. When Kara interviewed 700 people, she found 28% of them had a fear of death. How many of the people Kara interviewed suffer from necrophobia?

12 This pie graph shows the percentages of each blood type in the Australian population of 24 835 000. Calculate the number of people with each blood type.

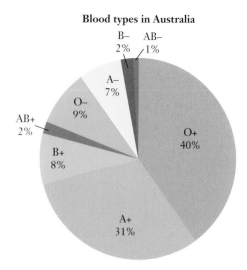

Blood types in Australia

INVESTIGATION

BLOOD TYPES AROUND THE WORLD

Is the percentage of people with the different blood types the same in all countries and continents?

What you have to do

Choose a country from each region below and research the percentages of people in the country that have each blood type.

– Europe	– South America
– Asia	– Africa
– North America	– Middle East

Copy and complete the table below to summarise your results. For each country, write in the percentages of each blood type. Add some extra countries of your choice if you like.

Country	A+	B+	AB+	O+	A–	B–	AB–	O–

Write a paragraph describing the similarities and differences in the percentages of blood types between different countries. How do other countries compare with Australia?

7.02 Comparing quantities using percentages

$$\text{Percentage} = \frac{\text{amount}}{\text{whole amount}} \times 100\% \quad \text{OR}$$
$$= \text{amount} \div \text{whole amount} \times 100\%$$

EXAMPLE 2

Express a test mark of 27 out of 45 as a percentage.

Solution

Write the result as a fraction and multiply by 100%. $\frac{27}{45} \times 100\% = 60\%$

Sometimes we have to change the units so they are the same before we can find the percentage.

EXAMPLE 3

What percentage of 2 days is 12 hours? ← If you are having trouble knowing which amount to write on the bottom of the fraction, look for the number following the 'of'.

Solution

Change 2 days into hours.	$2 \text{ days} = 2 \times 24 \text{ hours}$
	$= 48 \text{ hours}$
Write as a fraction and multiply by 100%.	$\frac{12}{48} \times 100 = 25\%$
Write your answer.	12 hours is 25% of 2 days.

Exercise 7.02 Comparing quantities using percentages

1 Express each test mark as a percentage.

   a   17 out of 20   **b**   51 out of 60

   c   14 out of 35   **d**   105 out of 140

2 Eliza scored the following test marks for her Year 11 exams.

   English: 43 out of 80   Science: 26 out of 40

   Mathematics: 69 out of 85   Business Studies: 59 out of 70

   Design: 38 out of 65   Health: 34 out of 45

   a   Calculate each test mark as a percentage. Give your answers correct to 2 decimal places.

   b   Rank Eliza's achievement in subjects from highest to lowest.

3 Over the last 3 basketball matches Tiago was successful with 18 shots out of 33 attempts, while Luke was successful with 23 shots out of 37 attempts.

   Who was more successful? Justify your answer.

4 Which food has the higher percentage of sugar in it?

   •   Muesli bars with 11 g of sugar in a 35 g bar   or

   •   Jam with 9.8 g of sugar in a 20 g serving

5 Colombia has an area of 1 141 748 km^2, of which 100 210 km^2 is water.

   Iran has an area of 1 648 195 km^2, of which 116 600 km^2 is water.

   a   Calculate the percentage of water for each country. Give your answer correct to 2 decimal places.

   b   Which country has a higher percentage of water?

   c   Why might the percentage of water be important?

6 Write the first quantity as a percentage of the second. Remember to express both quantities in the same units first.

   a   3 minutes, 1 hour   **b**   75 mm, 20 cm

   c   400 g, 32 kg   **d**   18 hours, 3 days

7 Jo exercises at the local gym 3 times a week. The table shows how much time Jo spent swimming and her total exercise time for her 3 visits last week.

	Swimming	Total exercise time
Monday	20 minutes	1 hour 10 minutes
Wednesday	35 minutes	1 hours 40 minutes
Saturday	45 minutes	2 hours 30 minutes

a Calculate the percentage of exercise time spent swimming on each visit. Give your answers correct to one decimal place.

b On which day did Joanne spend the highest percentage of her time swimming?

c Calculate the percentage of exercise time spent swimming for the week.

8 Which of the following cocktails has the higher percentage of alcohol?

A 750 mL of alcohol in a jug of 3 L   or

B 320 mL of alcohol in a bottle of 1.2 L

9 Two Australian batsmen score at the following rates:

- Clarkson hit 82 runs off 68 balls
- Bailden hit 48 runs off 38 balls

a Calculate the strike rate for each batsman, correct to 1 decimal place.

$$\text{Strike rate} = \frac{\text{Runs}}{\text{Balls}} \times 100$$

b Who is the better batsman based on these figures?

10 Which state has the higher rate of stamp duty payable on new cars?

- Victoria, $1230 stamp duty on a car purchase of $41 000
- Western Australia, $522.50 stamp duty on a car purchase of $19 000

PS

7. Applying percentages

STAMP DUTY WHEN BUYING A HOUSE

When you buy a house, you pay the state government a tax called stamp duty. In this investigation, you will find out the different rates for stamp duty in Australia.

What you have to do

- Search the Internet for an online calculator for stamp duty for house purchases.

- Suppose you are buying a $300 000 home and you are not a first home buyer. Use the online calculator to find the amount of stamp duty paid for each state/territory in Australia. Write this information down.

- Calculate the percentage rate used in ← $\text{Rate} = \dfrac{\text{stamp duty}}{\text{home price}} \times 100\%$
each state/territory.

- Which state/territory has the lowest stamp duty? Which has the highest?

- What is the difference between the lowest and the highest stamp duty?

- Why do you think the states and territories charge different amounts of stamp duty?

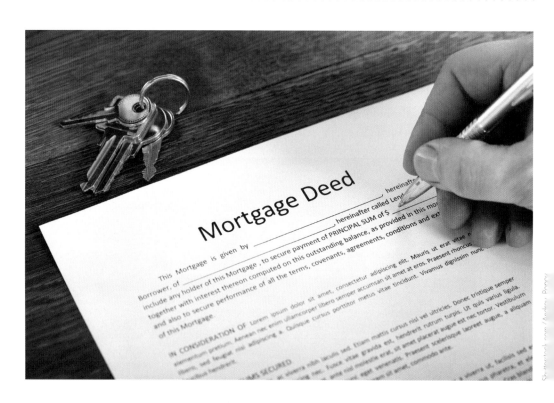

7.03 Simple interest

Interest is the money we earn when we invest with a bank, credit union or other financial institution. It can also be extra money paid back when we take out a loan. With **simple interest** (or **flat rate** interest), the interest earned or paid is a percentage of the **principal**, the original amount invested or borrowed.

WS

Simple interest 1

Simple interest formula

$$I = Pin$$

Sometimes this formula is written $I = Prn$.

where I = interest earned or paid

P = principal (what you invest or borrow)

i = interest rate per time period, as a decimal

n = number of time periods

What's the interest?

Simple interest riddle

EXAMPLE 4

Maurice invested $2000 at 3.25% p.a. simple interest for 3 years.

p.a. = per annum = per year. The interest rate is for a whole year.

a   How much interest will he earn?

b   How much will he have altogether at the end of 3 years?

Solution

a   In the simple interest formula $I = Pin$, $P = 2000$, $i = 3.25\% = 0.0325$, $n = 3$.

$I = 2000 \times 0.0325 \times 3$

$= 195$

Maurice will earn $195 in simple interest.

b   At the end of 3 years, Maurice will have his original investment plus the interest.

Total = $2000 + $195

$= $2195

Maurice will have $2195 altogether.

Sometimes money is not invested or borrowed for a whole number of years. We need to make sure the interest rate and the time period match!

EXAMPLE 5

Ruby borrowed $5600 at 3.9% p.a. for 18 months. How much interest did she have to pay?

Solution

When the interest rate is p.a. (years), the time must also be in years. Change 18 months to years by dividing by 12.

$I = 5600 \times 0.039 \times \dfrac{18}{12}$

$= 327.6$

In the simple interest formula $I = Pin$,

$P = 5600$, $i = 0.039$, $n = \dfrac{18}{12}$ years

Ruby paid $327.60 interest.

We can change the interest rate from a yearly rate (p.a.) to a rate for any other time period.

Divide the annual rate by:

- 52 for a weekly rate
- 26 for a fortnightly rate
- 12 for a monthly rate
- 4 for a quarterly rate (3 months)
- 2 for a six-monthly rate

EXAMPLE 6

What percentage interest rate per month is equivalent to 5.16% p.a.?

Solution

Divide the annual rate by 12 to change it to a monthly rate.

$5.16\% \div 12 = 0.43\%$

5.16% p.a. is equivalent to 0.43% per month.

Exercise 7.03 Simple interest

1 Calculate the simple interest for each investment.

a $2400 at 5% p.a. for 6 years   **b** $750 at 4% p.a. for 2 years

c $700 at 6.2% p.a. for 3 years   **d** $520 at 4.2% p.a. for 5 years

e $4200 at 7% p.a. for $3\frac{1}{2}$ years   **f** $3400 at 4.85% p.a. for $2\frac{1}{2}$ years

g $750 at 3.1% p.a. for $1\frac{1}{2}$ years   **h** $14 900 at 2.4% p.a. for 5.75 years

2 Nicole borrowed $2400 from a finance company for 2 years at 18% p.a. interest.

   a How much interest did she have to pay?

   b How much, including interest, did Nicole have to repay the finance company?

3 Carlos invested $1500 in an account paying 4.5% for 3 years.

   a How much interest did he earn?

   b How much, including interest, did Carlos have in his account at the end of 3 years?

4 What is the simple interest on each investment?

> Remember to make the units match! The p.a. in the interest rate means the time must be in years.

   a $1080 at 5% p.a. for 16 months

   b $475 at 2% p.a. for 5 months

   c $940 at 3.25% p.a. for 18 months

   d $4700 at 4.7% p.a. for 17 months

   e $12 560 at 7.05% p.a. for 6 months

   f $1560 at $4\frac{1}{2}$% p.a. for 3 months

5 Rodney won $500 as part of his prize for being the 'Apprentice of the Year'. He invested it for 1 month at 4.5% p.a. interest. How much does he have after 1 month?

6 What percentage interest rate per month is equivalent to 4.68% p.a.?

7 The NQM Bank offers investors 8.4% p.a. simple interest. Express, correct to 4 decimal places (where necessary), this rate of interest as a:

   a monthly rate   **b** weekly rate   **c** 6-monthly rate

   d daily rate   **e** fortnightly rate   **f** quarterly rate (3-monthly rate)

8 Yolanda invests $3000 with the NQM Bank using their 8.4% p.a. simple interest offer. Find the interest she earns if she invests her money for:

   a 4 months   **b** 21 weeks   **c** 6 months   **d** 45 days

   e 32 weeks (equal to 16 fortnights)   **f** 9 months (equal to 3 quarters)

9 Jeremy owed $365 on his credit card. The credit card company charged him one month's interest at 22% p.a.

   a How much interest was he charged?

   b Calculate the total amount he had to repay the credit card company.

7.04 Simple interest problems

Simple interest is also called flat rate interest because, unlike compound interest, the same amount of interest is paid or earned each year or time period.

Simple Interest 2

Simple interest

EXAMPLE 7

Melinda borrowed $24 500 for 3 years from NelsonNet Finance to buy a car. The interest rate is 9.5% p.a. flat.

  a How much interest will Melinda pay?

  b How much will Melinda pay the finance company to repay the loan?

  c Melinda is going to repay the loan over 5 years. How much will she have to pay each month?

Solution

  a $P = 24\,500$, $i = 0.095$, $n = 3$

$I = 24\,500 \times 0.095 \times 3$

$= 6982.5$

Melinda will pay $6982.50 interest.

  b Melinda has to repay the amount she borrowed plus the interest.

$24\,500 + \$6982.50 = \$31\,482.50$

Melinda will pay $31 482.50 to repay the loan.

  c First calculate how many months there are in 5 years by multiplying by 12.

$5 \times 12 = 60$ months

The total amount must be repaid over 60 monthly payments.

$\$31\,482.50 \div 60 = \$524.708\,33...$

$\approx \$524.71$

Melinda has to pay $524.71 each month.

To ensure that the loan is paid off completely, round *up* to the nearest cent when calculating the size of each payment.

NELSON QMATHS 11. Essential Mathematics   ISBN 9780170412650

Exercise 7.04 Simple interest problems

1 Vanessa borrowed $30 000 from a finance company to set up a catering business. She agreed to repay the money in monthly instalments over 5 years at 12% p.a. flat rate.

    a   How much interest will Vanessa have to pay?

    b   Calculate the total amount she must repay.

    c   How much must Vanessa repay each month?

2 Samir has just finished his apprenticeship and wants to borrow $1500 to buy tools. The bank offers him the money over 2 years at 8.7% p.a. flat rate interest, provided he makes monthly repayments.

    a   How much interest will Samir be charged?

    b   How much will Samir owe the bank company altogether?

    c   Calculate the value of his monthly repayments.

3 The Great Aussie Credit Union pays higher interest rates for large investments.

Balance	Flat interest rate
$1–$4999	3.00% p.a.
$5000–$9999	3.50% p.a.
$10 000–$19 999	4.00% p.a.
$20 000–$49 999	4.75% p.a.
$50 000 and over	5.50% p.a.

    a   Donna invested the $38 000 she inherited from her great-aunt with the credit union for 2 years. How much interest will she earn?

    b   Goran won $54 000 on lotto. He invested his win with the credit union for 18 months.

       i   How much interest will he earn?

       ii   How much money will be in the account at the end of the 18 months?

       iii   He decides to live on this money for the next 2 years. How much per month will he have to live on?

    c   Lauren saved $5750 from her part-time job at the supermarket.

       i   How much interest will she earn if she invests it with the credit union for 9 months?

       ii   Will she able to afford a holiday costing $5900? Justify your answer.

4 Sophie signed the contract to buy a new house before she sold her apartment. Six weeks later she needed to borrow $250 000 urgently to pay for the house as she hadn't sold her apartment. Banks have 'bridging finance' for this purpose. Sophie borrowed the money for 6 months at 15% p.a. Calculate the interest Sophie had to pay.

5 Angelo and Trish need $650 000 bridging finance for 3 months. The bank's interest rate for this is 11.5% p.a.

    a Calculate the interest they will have to pay.

    b Find the total amount they will have to repay.

    c During the three months they sell their current home for $575 000. How much will they still owe the bank?

6 Denise wants to buy a ride-on mower for her lawn-mowing business. The mower costs $9495 and she is offered a $700 trade-in on her old mower. The store offers her terms over 3 years at 7.4% p.a. flat rate interest after a deposit of $2500.

    a How much will Denise owe after she trades in her old mower?

    b How much will she owe after she pays the deposit?

    c How much interest will she be charged?

    d Calculate the total amount Denise must repay in instalments.

    e What is the amount of each monthly repayment?

7 When Jonny borrowed $18 000 from a finance company, he was charged 9% p.a. simple interest.

    a How much interest would he pay in one year?

    b His total interest bill was $6480. For how long did Jonny borrow the money?

8 When Yusuf borrowed $15 000, he was charged 7.8% p.a. simple interest. Including interest, Yusuf repaid $20 850. For how many years did Yusuf borrow the money?

9 Vanessa borrowed $12 000 from a finance company for 2 years. Including interest, she repaid the company $14 400.

    a How much interest did Vanessa pay per year?

    b Calculate the rate of simple interest p.a. the company charged her.

$$\text{Rate} = \frac{\text{Interest for one year}}{\text{Amount borrowed}} \times 100\%$$

10 Calculate the flat rate of interest p.a. being charged on each loan, using the same method as you used in question 9.

	Amount borrowed	Amount repaid (including interest)	Term (Length of the loan)
a	$30 000	$37 650	3 years
b	$14 600	$16 628	2 years
c	$12 300	$15 744	5 years
d	$45 000	$56 340	3.5 years

Chapter problem

You've covered the skills required to solve the chapter problem. Can you solve it now?

INVESTIGATION

SIMPLE INTEREST INVESTMENTS

We can construct a spreadsheet to calculate the simple interest earned on investments.

	A	B	C	D	E	F	G
1	Simple interest calculations						
2							
3	Enter the principal in F3					3500	
4	Enter the annual interest as a decimal rate in F4					0.05	
5	Enter the length of the investment in years in F5					3	
6							
7							
8		Interest earned =		525			
9		Total in the account =		4025			
10							
11							

1 Create a spreadsheet as shown above, except for the 2 values in column D.

2 What formulas should you enter in cells D8 and D9?

3 Enter these formulas and calculate the **Interest earned** and **Total in the account** for different values.

4 Change the values in column F to see how it affects the interest earned. Try small amounts and large amounts of money. Try low interest rates and high interest rates. Try a short time and a long time.

7.05 Percentage after percentage

In some situations we may need to calculate one percentage after another, for example, buying something on sale with an additional discount for cash or for paying your bill in a certain time. In this situation, we need to remember to do each percentage calculation separately.

Joanna purchases a new dress priced at $189 during a '20% off' sale.

a   How much discount does she receive on her dress?

b   What price does Joanna pay for the dress?

c   At the checkout, the shop assistant says there is an additional 5% off the discounted price if she pays cash. How much would this additional discount be?

d   What is the final price of the dress if Joanna pays cash?

e   Is this equal to 25% off the original price? Justify your answer with a calculation.

Solution

a   Calculate 20% of $189.

$$\text{Discount} = 0.02 \times \$189$$
$$= \$37.80$$

b   Discount price = original price − discount

$$\text{Discount price} = \$189 - \$37.80$$
$$= \$151.20$$

c   Calculate 5% of $151.20.

$$\text{Discount} = 0.05 \times \$151.20$$
$$= \$7.56$$

The additional discount is $7.56.

d   Final price = discount price − discount

$$\text{Final price} = \$151.20 - \$7.56$$
$$= \$143.64$$

The final price of the dress is $143.64.

e   Calculate 25% of $189.

$$0.25 \times \$189 = 47.25$$

Subtract this discount from the original price and compare with $143.64.

$$\$189 - \$47.25 = \$141.75$$

No, $143.64 is not the same as a 25% discount. A 25% discount is larger.

 NELSON QMATHS 11. Essential Mathematics   ISBN 9780170412650

We saw in Example **8** that we can't just add the percentages together and find a single percentage. However, we can find a single percentage that is equivalent to one percentage followed by another.

EXAMPLE 9

Dennis' Paints offers a 25% trade discount to builders and a further 5% discount if the account is paid within 10 days. Chris is a builder. He purchased $840 worth of paint and paid the account within 10 days.

a How much did he pay for the paint?

b What single discount is equivalent to a 25% discount followed by a 5% discount?

a First, calculate the trade discount.

Trade discount = 25% × $840 = $210

Then calculate the discounted price.

Trade price = $840 − $210 = $630

Then calculate the further discount.

Further discount = 5% × $630

= $31.50

Calculate the final price.

Final price = $630 − $31.50

= $598.50

b Find the total discount.

Total discount = $210 + $31.50

= $241.50

The percentage discount is $\frac{\text{total discount}}{\text{original price}} \times 100\%$

Percentage discount = $\dfrac{\$241.50}{\$840} \times 100$

= 28.75%

Exercise 7.05 Percentage after percentage

1 Brody buys a dinner suit costing $720 for his formal. The suit shop has a special formal price of 15% off and offers a further 5% off if Brody pays in cash.

   a How much discount does Brody receive for the special formal price?

   b What price does he pay for the suit?

   c Brody decides to pay cash for his suit. By how much more would his suit be discounted?

   d What is the final price of the suit?

   e Is this equal to 20% off the original price? Justify your answer with a calculation.

2 The advertised price for a Luxura car is $75 999. When Angelique visits the car dealership, she is offered an end-of-financial-year deal of 30% discount. If she finalises the deal within 2 days, she is offered an additional 5% discount.

   a What is the price of the Luxura car after the 30% discount?

   b What price would Angelique pay if she finalises the deal in two days?

   c Is this equal to 35% off the original price? Justify your answer with a calculation.

3 NQM News surveyed 2000 people about phone usage while driving. 58% of people interviewed admitted to texting or talking on the phone while driving. Of these people, 30% believed it was safe to do so.

   a How many people admitted to using the phone while driving?

   b How many people thought it was safe to do?

   c What percentage of the original sample thought it was safe to use the phone while driving?

4 Mainline Electrical offers a 20% trade discount to electricians and a further 5% discount for bills paid by the end of the month. Tina is an electrician. She buys $345 worth of electrical parts and pays her bill before the end of the month.

   a How much did she pay for the electrical parts?

   b What single discount is equivalent to a 20% discount followed by a 5% discount?

5 The Complete Discount Store is having an end-of-year sale, offering 25% off normal prices and a further 10% discount for cash.

   a Simon and Maddy plan to pay cash for a bed that normally sells for $2300. How much will they pay during the sale?

   b How much is their total saving?

   c What single percentage discount is equivalent to a 25% discount followed by a 10% discount?

6 On Mad Monday, Sam's Cameras offers 20% off everything in the store plus a further 5% discount on all PAINT brand cameras.

   a Find Sam's Mad Monday price for a PAINT camera usually priced at $730.

   b What single percentage would give the same discount to customers?

7 The Golden Camellia restaurant gives a 15% discount on all takeaway meals. They also give customers with a loyalty card a further 3% discount.

   a Kate and Jon are loyalty card customers. They order takeaway food that would cost $42 in the restaurant. How much did they pay for the meal?

   b What single percentage discount did they get? Give your answer correct to 1 decimal place.

8 Vision Discounts offers 20% off the cost price of TVs. However, it must then add 10% GST to the price.

   a What is the price of a 55 cm TV costing $800 after the discount?

   b What is the final price when the GST is added?

9 TV Deals 4 U add the GST to the cost of the TV and then offer a 20% discount.

   a What is the price of the same television as in Question 8 at TV Deals 4 U?

   b Which deal is cheaper for the customer – discount followed by GST or GST followed by discount? Justify your answer.

10 The population of Nelson Waters increased by 7% in 2017 and by 3.5% in 2018. At the beginning of 2017, the population was 2516. What was the population at the end of 2018?

PS

11 Judy and Keith deposit $10 000 in an investment account for 4 months at 5.5% p.a. interest. At the end of the 4 months they reinvest the $10 000 plus the interest they received for a further 4 months at 4.5% p.a. interest. How much interest have they earned from these 2 investments?

> To calculate the interest for 4 months, write it as a fraction of a year:
> 4 months = $\frac{4}{12}$ years.

PS

12 When a casting company advertised for extras to be part of a crowd scene for a movie, 12 400 people applied to be involved. The company rejected 42% of the applicants because they were too young and they rejected 65% of the remainder because they were too tall. How many of the applicants were neither too young nor too tall?

PS

7.06 Percentage problems

In this chapter, we have used various techniques for different types of problems involving percentages. Deciding which technique to use is an important skill.

$$\text{Percentage of a quantity} = \frac{\text{Percentage}}{100} \times \text{quantity} \qquad \text{OR}$$

$$= \text{Percentage} \div 100 \times \text{quantity}$$

$$\text{Finding a percentage} = \frac{\text{amount}}{\text{whole amount}} \times 100\% \qquad \text{OR}$$

$$= \text{amount} \div \text{whole amount} \times 100\%$$

$$\text{Simple interest: } I = Pin$$

For each problem in the following exercise, decide which strategy you will use before you start the question. You may like to use the problem-solving stages for these problems.

Exercise 7.06 Percentage problems

1 Cindy booked an overseas holiday costing $3650. The travel agent offered her a 17% discount if she paid 90 days in advance.

   a How much will Cindy save if she takes advantage of this offer?

   b Cindy decides to pay 90 days in advance. How much will the trip cost her?

2 Damien borrowed $16 000 from a finance company to buy a new car. He was charged 13% p.a. flat rate interest and agreed to pay the loan back over 5 years.

   a How much interest was Damien charged by the company?

   b How much did he repay in total?

   c Damien repaid the loan by making monthly payments. How much did he have to pay each month?

3 Gabrielle owns a fashion shop. When items come into the store, Gabrielle increases the wholesale price by 95%, and then adds another 10% for GST to determine the final selling price.

   a Gabrielle receives a dress costing $140. Calculate the price of the dress without GST.

   b Determine the price of the dress including GST.

   c What price do you think Gabrielle will put on the price tag? Give reasons for your answer.

   d The wholesale price of a suit is $125. What amount, including GST, will Gabrielle write on the price tag?

4 Thomson Valley Council will not approve house plans if the floor area is greater than 40% of the area of the land on which it is to be built. Jackie's block of land is 480 m². What is the floor area of the largest house the council will allow her to build on her block of land?

5 During the summer sales Felicity bought a towel that is normally priced at $48 for $35. What percentage discount did she receive? Answer to the nearest whole percentage.

6 Holly's new lounge cost $3000 and it is depreciating (decreasing) at a rate of 10% per year.

   a How much will the lounge be worth at the end of 1 year?

   b Find the value of Holly's lounge when it is 2 years old.

   c Is this the same as depreciating by 20% of the original price? Justify your answer with a calculation.

iStock.com/PeopleImages

7 Xander owns 700 shares in the company Telco. Each share has a market value of $9.50.

   a How much are Xander's shares worth?

   b Telco pays a dividend of 5% of the value of the shares. How much does Xander receive as his dividend?

8 Ken is charged $65 for a visit to the doctor. The Medicare rebate given to Ken is $35.60. What percentage of the doctor's charge is the Medicare rebate? Answer correct to 1 decimal place.

9 Forster Auto Superstore gives customers who pay cash a 5% discount, but it charges customers who pay with a credit card an additional 1.5%. The shop advertised a set of car mats for $60.

   a What is the 'cash price' of the mats?

   b How much will Joe, who pays with a credit card, be charged for the mats?

10 TVs R Us advertised a TV for $3999. It can be bought with 10% deposit and weekly repayments of $38.75 over two years.

   a   Marco bought the TV this way. How much deposit does he pay?

   b   How much does Marco owe after paying the deposit?

   c   Calculate the amount Marco pays in the weekly repayments.

   d   How much does Marco pay altogether?

   e   How much more than the sale price did Marco pay?

   f   The extra calculated in part **e** is the interest Marco is charged. How much interest was Marco charged for ONE year?

   g   What rate of interest was Marco charged?

$$\text{Interest rate} = \frac{\text{interest charged for 1 year}}{\text{amount owing after deposit is paid}} \times 100\%$$

11 Yvonne owns Bucknall Mining shares valued at $15.20 each. The company pays a dividend of 53 cents per share. What percentage was this dividend of the value of the share? Give your answer correct to 2 decimal places.

> Remember: units in each quantity must be the same!

Percentages
find-a-word

KEYWORD ACTIVITY

deposit	depreciate	discount	dividend
flat rate interest	GST	interest	per annum
principal	repayment	simple interest	stamp duty

   1   For each word in the list, write a definition or explanation of the word and then use it in a sentence showing its meaning.

   2   Read over the exercises in the chapter and write down any words that you are not familiar with. Write a sentence explaining the meaning.

SOLUTION ᵀᴼ ᴛʜᴇ CHAPTER PROBLEM

Problem

Aron buys some bedroom furniture on a store finance arrangement. The bedroom furniture costs $3200. The store finance will charge him 17% p.a. interest and expects him to pay it off monthly over 3 years. How much will Aron have to pay each month?

Solution

WHAT?

STAGE 1: WHAT IS THE PROBLEM? WHAT DO WE KNOW?

We have to work out Aron's monthly repayment.
Cost is $3200, interest rate is 17% p.a., monthly repayments over 3 years.

SOLVE

STAGE 2: SOLVE THE PROBLEM

We break this problem down into smaller steps:

Calculate the simple interest: Interest = Pin
$$= \$3200 \times 0.17 \times 3$$
$$= \$1632$$

Calculate the total owed: Total repaid = $3200 + $1632
$$= \$4832$$

Calculate the cost per month: In 3 years there are $3 \times 12 = 36$ months.
Payment each month = $4832 ÷ 36
$$\approx \$134.23 \text{ (rounding up)}$$

CHECK

STAGE 3: CHECK THE SOLUTION

Check: $134.23 × 36 = $4832.28 (which is over $4832).
$134.23 seems a reasonable amount to pay off per month (not too high).

PRESENT

STAGE 4: PRESENT THE SOLUTION
Aron will have to pay $134.23 per month.

Applying percentages

1 **a** Mark's pay is $2764.80 per fortnight. He is saving 15% of his pay to buy a new car. How much does he save each fortnight?

b There are 135 students in Year 12 at Nelson Waters High. Approximately 78% of Year 12 have paid for their formal tickets. How many students have paid for their tickets?

2 **a** When the Jets netball team played on the weekend, Megan shot 23 goals out of 31 attempts and Jasmine shot 33 goals out of 48 attempts.

    i Calculate the success rate for each shooter as a percentage. Give your answer correct to 2 decimal places.

    ii Who was the better shooter on the weekend?

b The Brisbane Broncos played the North Queensland Cowboys in an NRL match. The Broncos completed 17 out of 22 tackle sets. The Cowboys completed 19 out of 24 tackle sets.

    i Write each completion rate as a percentage. Give your answer correct to the nearest whole number.

    ii Who achieved the better completion rate?

3 **a** Calculate the simple interest earned on $5100 invested at 3.5% p.a. for 4 years.

b Priya borrowed $13 000 at 7.95% p.a. for 3 years.

    i Calculate the interest Priya will be charged.

    ii What is the total amount she will have to repay?

4 **a** What is the simple interest payable on $2000 borrowed at 4.25% p.a. for 18 months?

b Janine sells her car for $17 500 and wants to invest the money for 5 months to finance her future travel. The Advantage Bank offers 3.15% p.a. in an investment account. How much interest will Janine earn?

NQM Finance charges 12.5% p.a. simple interest on short term loans. Use this information to answer questions **5** and **6**.

5 Express (correct to 4 decimal places where necessary) this rate of interest as a:

    a monthly rate   **b** weekly rate   **c** 6-monthly rate

6 Neil borrows $7000 from NQM Finance. Find the interest he pays if he borrows the money for:

    a 5 months   **b** 17 weeks   **c** 6 months

7 Madeline borrows $6300 to set up a small business. She is charged 5.2% p.a. simple interest and she takes out the loan for 3 years, repaying what she owes by making equal monthly repayments.

   a How much interest will Madeline have to pay?

   b Calculate the total amount she must repay.

   c How much will Madeline have to pay each month?

8 Vamsee buys a new Volkswagen. The price is $39 900 with a 'model runout' sale discount of 22%. If he buys a display car already in the saleyard he receives a further 5% discount.

   a What is the price of the car after the sale discount?

   b What price does Vamsee pay if he buys a display car?

9 Franken Furniture is having a mid-year sale. All items in the store are 20% off normal prices. Customers who pay in cash are offered a further 10% discount.

   a Joshua and James pay cash for a lounge suite that normally sells for $2599. How much will they pay?

   b How much is their total saving?

   c What single percentage discount is equivalent to a 20% discount followed by a 10% discount?

10 Manuel is charged $304.60 for X-rays of his foot. The Medicare rebate for Manuel is $164.60. What percentage of the charge is the Medicare rebate? Answer correct to 1 decimal place.

11 For tax purposes, equipment bought for a business is depreciated at 20% per year. Stefan buys a photocopier for $3070.

   a Calculate the amount of the depreciation in the first year.

   b What is the photocopier valued at after 1 year?

   c Find the value of the photocopier at the end of the second year.

REPRESENTING DATA

8.

SHOW ME THE DATA

Chapter problem

The Nelsonville College principal thought that too many cars were speeding through the school zone after school when the speed limit is 40 km/h. One day, the police recorded the speeds of all cars for 1 hour after school.

What type of data is this? What would be the best way to organise and present this data?

WHAT WILL WE DO IN THIS CHAPTER?

- Distinguish between categorical data and numerical data
- Graph categorical data and numerical data
- Graph and interpret histograms, dot plots, stem-and-leaf plots and choose the best display for a particular set of data
- Identify outliers in a set of data

HOW ARE WE EVER GOING TO USE THIS?

- When we need to organise information we have collected
- When preparing a presentation of information for a group of people
- When evaluating data presented in the media

8.01 Categorical and numerical data

A collection of facts or information is called **data**. There are two main types of data.

Statistical data match-up

> **Categorical data** is information that is grouped into categories, such as the colour of cars or ice cream flavour.
>
> **Numerical data** is information that is counted or measured as numbers, such as the number of goals scored or a person's height.

EXAMPLE 1

Classify each type of data as categorical or numerical.

a makes of cars   **b** people's salaries

c weights of athletes   **d** favourite radio stations

Solution

a	Makes of cars are words.	Categorical
b	Salaries involve numbers.	Numerical
c	Weights of athletes involve measurement on a numerical scale.	Numerical
d	Radio stations involve words.	Categorical

Exercise 8.01 Categorical and numerical data

1 Classify each type of data as categorical (C) or numerical (N).

  a masses of Year 11 students   **e** monthly rainfall in Mount Isa

  b brands of computers   **f** causes of tooth decay

  c prices of large pizzas   **g** sets won in a tennis match

  d number of TVs owned by families   **h** speed of trucks

2 Classify each type of data as categorical (C) or numerical (N).

  a people's hair colours   **e** countries where cars are manufactured

  b customer ratings of a restaurant's level of service (from poor to excellent)   **f** school populations

  c number of homes with smart TVs   **g** heights of bridges

  d times taken for athletes to run 100 metres   **h** classification ratings of films (for example, PG)

3 You are asked to do a survey on people's preferences about anything to do with cars. Write a questionnaire with 6 questions: 3 questions whose answers are categorical and 3 questions whose answers are numerical.

8.02 Displaying categorical data

When we collect categorical data, we can organise it into a frequency table. This makes it easier to see what the information is telling us.

Heidi asked a group of her fellow students which night of the week they would prefer to have the school ball. She recorded the responses as T (Thursday), F (Friday), Sa (Saturday) and Su (Sunday). This is her data.

F   F   T   Sa   F   Sa   F   Sa   Sa   F   Sa   T   Sa   F

F   T   T   Su   F   Sa   T   Sa   F   F   F   F   T   T

Arrange this data in a frequency table.

Solution

A frequency table has three columns. Complete the tally column by putting one tally mark for each response.

Day of the week	Tally	Frequency
Thursday	ⅢⅠ ⅠⅠ	
Friday	ⅢⅠ ⅢⅠ ⅠⅠ	
Saturday	ⅢⅠ ⅠⅠⅠ	
Sunday	Ⅰ	

Count the tally marks to complete the frequency column.

Day of the week	Tally	Frequency
Thursday	ⅢⅠ ⅠⅠ	7
Friday	ⅢⅠ ⅢⅠ ⅠⅠ	12
Saturday	ⅢⅠ ⅠⅠⅠ	8
Sunday	Ⅰ	1
Total		**28**

We can display categorical data in a column graph.

EXAMPLE 3

Display the data from Example 2 in a column graph.

Day of the week	Frequency
Thursday	7
Friday	12
Saturday	8
Sunday	1

Solution

The horizontal axis will be the days of the week.

The vertical axis will be the frequency.

Columns will have spaces between them.

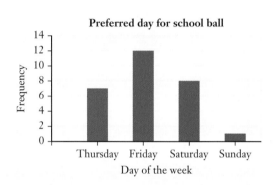

Exercise 8.02 Displaying categorical data

1 Piyush recorded the methods students in his year use to travel to school. He recorded the data as W = walk, R = ride a bike, B = bus, C = car and T = train. This is his data.

B B C T W	R W W T C	T T C B R	W W R R R
W W C R B	B W T W R	W W B R B	R W W T C
R R W W T	C W R W W	C B W W R	R R W B R
W R R W C	T R W W B	R R C B T	W R R B C

a Arrange this data in a frequency table.

b What is the most common way for students to travel to school?

c What is the least common travel method?

2 Tiana surveyed a selection of homes in Nelson Waters to find out the ages of each child. She recorded the data as B = babies up to 2 years, P = pre-schooler, I = early primary and O for children older than 8 years. This is Tina's data.

B B O I O   O B P P P   O B I I O   O I I I P   O B B O I

P O O P I   P P O B I   I O P O O   B I P B I

a Arrange Tiana's data in a frequency table.
b Which category of children contains the largest number of children?
c What percentage of children in the survey are babies?

3 Jasmine asked 30 students in Year 11 'What is your preferred brand of car?'

Her results are shown below and coded as follows: Holden = H, Ford = F, Toyota = T, Mitsubishi = M, Subaru = S, Other = O

H H T F F   F M T O S   T H H H F   S M T T H   F F H H O   S T M H F

a Arrange Jasmine's data in a frequency table.
b What is the most popular brand of car amongst these students?
c What percentage of students preferred Mitsubishi?

4 Tristian surveys a class of 30 students about the day of the week that each student's birthday falls on this year. The results are shown in the table below. Display this data in a column graph.

Example 3

Day	Monday	Tuesday	Wednesday	Thursday	Friday	Saturday	Sunday
Number of students	6	3	4	6	3	5	3

5 Allison counted the colours of the jelly beans in a large jar.

Colour	Red	Black	Purple	Pink	Orange	White
Frequency	110	50	20	40	60	80

Display this data as a column graph. ← You can use a spreadsheet to construct the graph if you want to!

8.03 Displaying numerical data

We can also summarise numerical data into a frequency table and present it in a graph.

EXAMPLE 4

Kyle asked 25 students in his class: 'How many children are there in your family?' The answers are recorded below:

1, 6, 7, 4, 3, 3, 2, 2, 5, 3, 3, 3, 3, 1, 7, 3, 2, 3, 2, 5, 4, 4, 1, 1, 2

Arrange Kyle's information in a frequency table.

Solution

Number of children	Tally	Frequency
1	IIII	4
2	IIII	5
3	IIII III	8
4	III	3
5	II	2
6	I	1
7	II	2

We often graph numerical data using a **histogram**. A histogram is like a column graph but there are no spaces between the columns. We show frequency on the vertical axis and we show the scores on the horizontal axis.

Shutterstock.com/goodluz

EXAMPLE 5

Draw a histogram for the data in Example 4.

Solution

Place 'Number of children' on the horizontal axis.

Place 'Frequency' on the vertical axis.

Make sure that the columns are centred so that the values on the horizontal axis are directly below them.

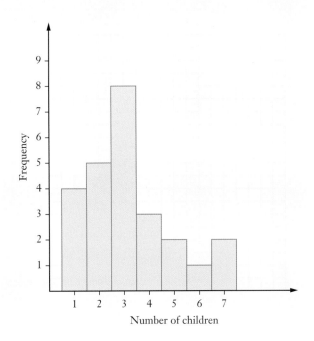

Exercise 8.03 Displaying numerical data

1 The number of hamburgers sold each day in a food court between 12 noon and 2 p.m. in August is listed below.

17   27   28   18   18   17   19   19   25   27   17   19   20   19   21   26

28   18   19   20   17   19   23   24   20   18   17   20   19   27   28

   Example **4**

a Arrange this information in a frequency table.

b On how many days did the shop sell fewer than 20 hamburgers?

c On what percentage of days did it sell more than 25 hamburgers? Round your answer to the nearest whole number.

Shutterstock.com/Sorbis

2 Sajjid asked a group of Year 11 students how many siblings (brothers and sisters) they had. The results were as follows:

4 0 2 3 2 4 3 4 3 0 0 1 2 3 2 2 1 4 3

1 1 1 2 0 1 2 1 0 1 3 2 3 4 1 1 3 3

    a   Complete a frequency table for Sajjid's data.

    b   How many students were asked this question?

    c   How many students had no siblings?

    d   How many students had fewer than 3 siblings?

    e   What percentage of students had 1 sibling? Round your answer to the nearest whole number.

3 A life insurance company recorded the ages of 40 clients.

84 76 39 45 38 74 66 81 49 57

59 42 31 43 71 80 40 37 73 87

77 49 53 62 62 74 84 90 31 47

52 62 79 33 31 52 46 43 55 38

    a   Copy and complete the grouped frequency table below for this data.

Age	Tally	Frequency
31–40		
41–50		
51–60		
61–70		
71–80		
81–90		
Total		

    b   Which age group has the most clients in it?

    c   How many clients were 50 years old or younger?

    d   What percentage of clients were over 70 years old?

Example 5

4 Amber surveyed 20 households in her street about how many phones they owned.
The results are shown in this frequency table.

Draw a histogram to display Amber's data.

Number of phones	Frequency
0	2
1	7
2	8
3	2
4	1

5 This frequency table shows the number of letters received by an office each day.

Number of letters	Frequency
3	4
4	7
5	8
6	14
7	12
More than 7	10

a   Draw a histogram for this data.

b   On how many days did the office receive fewer than 5 letters?

c   Which number of letters had the highest frequency?

6 Atiq counted the number of phone calls he made each day for 30 days.

5   3   6   2   4   3   3   5   4   7   5   1   6   3   2

3   5   6   1   6   2   4   3   1   5   4   2   7   3   2

a   Complete a frequency table for this data.

b   Draw a histogram for this data.

c   On how many days did Atiq make 5 calls?

d   What was the most common number of phone calls made each day?

7 The P.E. department recorded the heights of Year 11 students in centimetres.

151   167   181   172   179   155   159   162   169   174   178   180   158

166   171   168   157   160   175   172   150   169   163   170   176

a   Copy and complete this grouped frequency table.

Class	Tally	Frequency
150–154		
155–159		
160–164		
165–169		
170–174		
175–179		
180–184		
Total		

b   How many students had their height recorded?

c   Draw a histogram for this data.

d   How many students were 170 cm or taller?

e   Which class interval has the most students in it?

8. Show me the data

8.04 Dot plots and stem-and-leaf plots

Dot plots are used for small sets of numerical data that are close together.

Stem-and-leaf plots are used for larger sets of numerical data and list the actual data values.

EXAMPLE 6

The daily maximum temperature (in °C) in Cairns in July was recorded for 15 days.

32 30 31 32 31 30 31 31 31 31 29 25 28 27 29

Construct a dot plot for this data.

Solution

The temperatures go from 25° to 32°, so the line for the dot plot should go from 24 to 33.

Add a dot above the line for each data value.

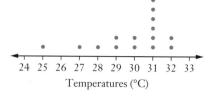

Temperatures (°C)

EXAMPLE 7

Sarah works for law enforcement and recorded the number of malicious property damage reports in the inner city each month over 3 years.

| 54 | 41 | 55 | 49 | 37 | 38 | 37 | 48 | 51 | 44 | 52 | 44 | 58 | 70 | 60 | 46 | 63 | 54 |
| 45 | 43 | 46 | 55 | 55 | 67 | 49 | 66 | 90 | 45 | 66 | 62 | 51 | 51 | 53 | 53 | 38 | 52 |

Draw an ordered stem-and-leaf plot for this data.

Solution

The stem will be the tens digits. The smallest number is 37 and the largest number is 90, so the stem will go from 3 to 9.

Stem	Leaf
3	
4	
5	
6	
7	
8	
9	

The leaf will be the units digit for each number in the data. This is called an unordered stem-and-leaf plot because the 'leaves' are not in order.

Stem	Leaf
3	7 8 7 8
4	1 9 8 4 4 6 5 3 6 9 5
5	4 5 1 2 8 4 5 5 1 1 3 3 2
6	0 3 7 6 6 2
7	0
8	
9	0

key: 3 | 7 = 37

Then we write the 'leaves' in order. This is called an ordered stem-and-leaf plot.

Stem	Leaf
3	7 7 8 8
4	1 3 4 4 5 5 6 6 8 9 9
5	1 1 1 2 2 3 3 4 4 5 5 5 8
6	0 2 3 6 6 7
7	0
8	
9	0

key: 3 | 7 = 37

Fairfax Media/Jeffrey Chan

Exercise 8.04 Dot plots and stem-and-leaf plots

1 Ahmed surveyed his maths class to find the number of hours each student spent on the Internet each week. These are the results.

14 15 17 18 13 16 19 19 18 14

17 15 13 13 14 19 13 14 17 18

  a Draw a dot plot for Ahmed's data.

  b How many students were in the class?

  c How many students used the Internet for 14 hours per week?

  d How many students used the Internet for more than 16 hours per week?

2 The number of motor vehicle thefts in an inner-city suburb was recorded each month for one year.

15 9 11 16 13 11 13 12 10 15 13 9

  a Draw a dot plot for this data.

  b In how many months were 13 motor vehicles stolen?

  c What percentage of months had 13 motor vehicles stolen?

  d In how many months were fewer than 12 motor vehicles stolen?

3 Lisa surveyed her group of friends about the amount of money they spent on fuel last week. The answers were rounded to the nearest dollar. This is the data she recorded.

$20 $28 $25 $26 $22 $26 $28 $28 $24 $22 $29 $28

  a Draw a dot plot for this data.

  b How many friends did Lisa survey?

  c What was the most common amount of money spent on fuel?

  d How many of Lisa's friends spent less than $26 on fuel?

4 Mrs White, the school canteen manager, records the daily number of students visiting the canteen over 3 weeks.

105 76 97 88 114 86 124 101 112 98 95 105 117 81 112

  a Show this information in a stem-and-leaf plot.

  b On what percentage of days were more than 100 students served? Answer correct to the nearest whole percentage.

5 A security firm recorded the monthly number of shoplifting incidents over 3 years.

20 20 23 11 12 33 22 30 16 17 35 48

25 27 25 34 20 23 25 17 12 14 13 13

48 42 55 33 24 39 26 41 33 31 19 55

   a Show this data on a stem-and-leaf plot.

   b How many months had the number of incidents in the 30s?

   c What percentage of the total months was this? Answer correct to the nearest whole percentage.

6 A class of Year 11 students were surveyed on the number of hours of part-time work they did last month.

42 16 35 27 9 0 33 21 14 11

26 29 31 22 8 24 5 0 15 25 17

   a Complete a stem-and-leaf plot to show this information.

   b What was the most number of hours worked by anyone in this sample?

   c How many students worked between 10 and 20 hours in the month?

   d How many students worked more than 20 hours in the month?

   e How many students were in this class?

   f What percentage of students worked more than 20 hours in the month? Answer correct to the nearest whole percentage.

7 Decide which graph – a histogram (H), a dot plot (D) or a stem-and-leaf plot (SL) – would be best to use for the given data. Justify your answer.

PS

   a Number of apples per tree

   b Minimum temperature each day for a fortnight in February in Bundaberg

   c Salaries of the 40 employees of a small business

   d Time taken to travel to school in minutes by students in Year 12

   e Rainfall each day for the month of August in Townsville

   f Number of mobile phones per family in Nelson St

   g Ages of patients in a doctor's practice

   h Sales of shirts at all branches of a fashion store in one week

   i Number of students in each class at St Judy's Primary School

   j Heights of all Year 11 students at Thomson Grammar School

Chapter problem

You've covered the skills required to solve the chapter problem. Can you solve it now?

8.05 Outliers

Sometimes with numerical data we have a number that is an **outlier**, a value that is very different from the rest of the data. It can be either much bigger or much smaller than the other scores. An outlier can make it difficult to graph the data. When we see a score that looks like an outlier, we should check that the score is reasonable because it could be a mistake.

EXAMPLE 8

Lochie asked eight friends about the amount of pocket money they received each week. The results were:

$20 $32 $32 $40 $18 $32 $18 $75

a   Identify the outlier in this data.

b   Is this outlier reasonable or likely to be a mistake?

Solution

a   Write the data in order.

18, 18, 20, 32, 32, 32, 40, 75

Choose the score that is either much bigger than the other scores or much smaller than the other scores.

The outlier here is $75 because it is much bigger than the other scores.

b

It is reasonable, as it is quite possible for one student to receive much more pocket money than other students.

Exercise 8.05 Outliers

Example 8

1 For each set of data, identify the outlier.

a	12	15	28	19	15	14	16	
b	7	5	6	8	7	1	8	6	9
c	32	35	12	40	36	29	38	30
d	94	49	35	38	31	44		

2 Data can have more than one outlier. What are the outliers in this data set?

6   8   0   6   8   16   8   7   2   8   9   9   8

3 The results of a mathematics test (out of 100) for a Year 10 class are shown.

55   52   50   45   55   45   60   50   60   58
75   45   49   59   58   59   56   49   31   52

a   Sort these results in order from highest to lowest.

b   What are the outliers in this data?

4 The nurse recorded the masses (in kg) of patients in the hospital emergency room.

68	59	63	80	68	54	48
49	64	47	48	59	68	30

a Place these results in order from highest to lowest.

b What are the outliers in this data set?

5 Katrina measured the heights of 25 Year 11 male students in centimetres.

175 176 185 176 25 184 197 161 186 169 171 170 182

165 179 180 167 169 198 167 170 180 182 173 230

a Draw a stem-and-leaf plot of this data.

b What are the outliers for this data?

c For each outlier, decide if it is reasonable or likely to be a wrongly-recorded height. Explain your answer.

6 The prices of 8 houses sold in Nelson Waters Rd over the last 2 years are:

$620 000	$700 500	$738 000	$625 000
$598 000	$696 500	$720 000	$1 800 000

a What is the outlier for this data?

b Is this outlier reasonable or is it likely to be a mistake? Explain your answer.

7 Ten people work in a small business which sells stationery. Their annual salaries are:

$61 000	$57 000	$66 000	$51 000	$53 000
$62 000	$56 000	$63 000	$60 000	$245 000

a What is the outlier for this data?

b Is this outlier reasonable or is it likely to be a mistake? Explain your answer.

8 For the same sample of Year 11 males used in question **5**, Katrina recorded the length of the right foot in centimetres. This is her data.

35	27	28	28	25	28	30	27	27	25	28	26	29
10	26	30	15	27	33	25	25	20	33	11	40	

a How many of these scores are in the 20s?

b What are the outliers in this data?

c For each outlier, decide if it is reasonable or likely to be a wrongly recorded foot length. Explain your answer.

9 Francine is a florist. In the first 8 years of her business, her annual profits were $35 000, $66 000, $69 000, $72 000, $75 000, $71 000, $68 000 and $76 000.

a What is the outlier for this data?

b Is this outlier reasonable or is it likely to be a mistake? Explain your answer.

DATA IN EVERYDAY LIFE

In this investigation you will find real-life examples of data presented in different ways. Here are some of the key words from this chapter:

categorical data   numerical data   histogram

dot plot   stem-and-leaf plot   outlier

What you have to do

- Search the Internet for each keyword/phrase, including images. You may like to search for Australian examples.

- Find an example for each one including a graph. Try to find one from real life, not from a maths site.

- Save it and explain its context and purpose.

- Create a PowerPoint presentation of what you have found.

Data
find-a-word

KEYWORD ACTIVITY

categorical data   numerical data   frequency table   histogram

dot plot   stem-and-leaf plot   outlier

Match each word above with one of the following definitions.

A   a table summarising data

B   a graph using dots to make columns

C   information that is grouped into categories, such as the colour of cars

D   a score that is very different from the rest of the data

E   looks like a column graph but there are no spaces between the columns

F   information that is counted or measured, such as the number of goals scored

G   graph used for larger data sets that keeps all the detail of the data

SOLUTION TO THE CHAPTER PROBLEM

Problem

The Nelsonville College principal thought that too many cars were speeding through the school zone after school when the speed limit is 40 km/h. One day, the police recorded the speeds of all cars for 1 hour after school.

What type of data is this? What would be the best way to organise and present this data?

Solution

WHAT?

STAGE 1: WHAT IS THE PROBLEM? WHAT DO WE KNOW?

We need to determine:

- what type of data we are collecting
- how to organise and present this data

SOLVE

STAGE 2: SOLVE THE PROBLEM

The data is speed, which are numbers, so the data is numerical.

Numerical data is best organised into a frequency table. If the speeds are spread over a wide range, it may be necessary to use a grouped frequency table.

Data is usually presented in a graph. For a few cars, we could use a dot plot, but as there will probably be a lot of cars, we should use a histogram or a stem-and-leaf plot.

As the principal is interested in the number of cars doing more than 40 km/h, it would be useful to calculate this number.

CHECK

STAGE 3: CHECK THE SOLUTION

We have answered all parts of the problem.

PRESENT

STAGE 4: PRESENT THE SOLUTION

The data is numerical. It should be organised into a frequency table and presented using a histogram or a stem-and-leaf plot.

8. CHAPTER REVIEW

Show me the data

Exercise 8.01

1 Classify each type of data as categorical (C) or numerical (N):

  a brands of toothpaste

  b hours of TV watched per week

  c number of employees in small businesses

  d country of birth of students at a school

  e heights of school students

  f ratings given by customers for a restaurant

Exercise 8.02

2 Marisa surveyed Year 11 students about their favourite colour from the following list:

  Blue (B)   Green (G)   Red (R)   Purple (P)   Yellow (Y)   Magenta (M)

  These are their results:

  BBPRB   GYYMG   BGRPY   MMYPP   RRGBG   PYRGM

  YPGRP   BBBBM   GGRPY   GGBBB   YYPPR   MMBBG

  a Arrange this data in a frequency table.

  b What is the most popular colour?

  c What percentage of students chose green?

  d Display this data in a column graph.

Exercise 8.03

3 The temperature at 11 a.m. each day is recorded:

  15°C   22°C   17°C   21°C   18°C   20°C   19°C   21°C

  20°C   19°C   22°C   21°C   21°C   19°C   15°C   21°C

  22°C   18°C   21°C   20°C   22°C

  a Complete a frequency table for this data.

  b For how many days was the temperature recorded?

  c How many days had temperatures below 20°C?

  d Draw a histogram for this data.

4 These are the results for Andre's Year 11 class for a quick maths quiz out of 20.

10	14	13	17	16	11	14	14	12	14	13	16

15	12	16	13	12	15	10	14	13	16	14

  a Draw a dot plot for Andre's data.

  b How many students in the class?

  c How many students scored more than 14 out of 20 for the quiz?

5 *Tech-to-go Gamestore* records the age of each customer who comes into the store one morning.

25	55	36	29	28	50	47	39	52	41	33

50	29	28	56	33	26	35	35	48	32

  a Show this information on a stem-and-leaf plot.

  b How many customers came into the store that morning?

  c What age was the oldest customer that morning?

  d What percentage of customers were under the age of 35? Answer correct to 1 decimal place.

6 For each set of data, identify the outlier.

  a 27   28   19   26   27   25   28   27

  b 0   3   2   3   1   7   1   0   0   2   3

  c 43.1   43.3   43.1   43.2   43.9   43.0   43.1   43.0

  d 95   101   98   76   97   105   103   99

7 NelsonNet Airlines flies from Brisbane to Sydney. The number of passengers on the 7 a.m. flight for 10 days was recorded.

56	100	98	104	125	101	89	93	100	99

  a Write these scores in order from lowest to highest.

  b What are the outliers in this data?

  c For each outlier, decide if it is reasonable or likely to be a mistake. Explain your answer.

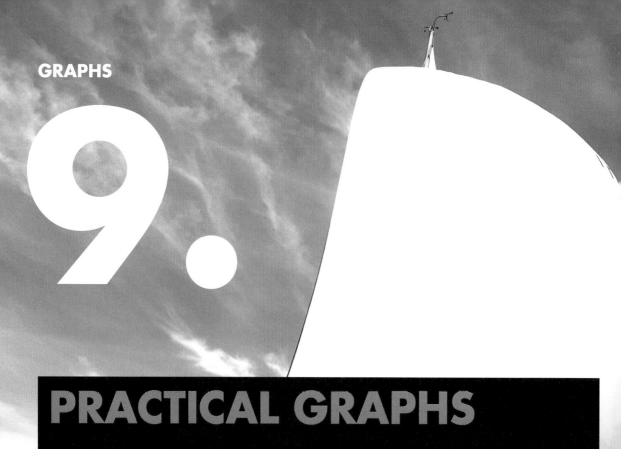

9.

PRACTICAL GRAPHS

Chapter problem

Paul owns a factory that makes sails. It costs $9000 per month to cover rent, electricity and wages in the factory. Each sail costs $250 to make and sells for $850.

a How many sails does Paul need to make and sell per month to break even?

b What profit or loss will he make in a month if he makes and sells 20 sails?

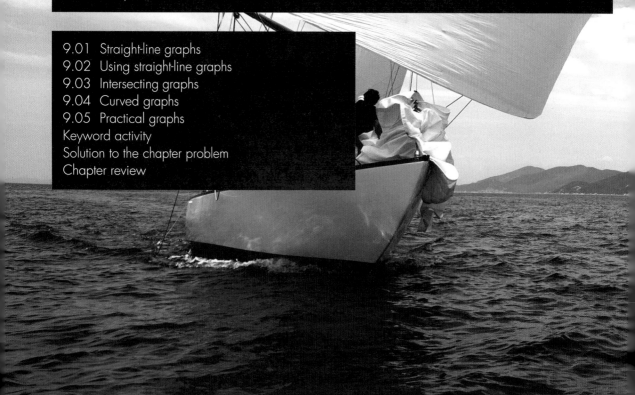

WHAT WILL WE DO IN THIS CHAPTER?

- Graph straight lines and find the gradient and y-intercept
- Use straight-line graphs in practical situations and interpret the meaning of the gradient and y-intercept
- Solve problems involving 2 intersecting lines and identify break-even points for a business to make a profit
- Graph curves and use curves in practical situations
- Draw a graph of a practical situation that doesn't follow a formula, such as the noise level of a classroom over a lesson period

HOW ARE WE EVER GOING TO USE THIS?

- When we want to represent relationships graphically
- When we model real-life situations with algebra and a graph

9.01 Straight-line graphs

We can graph algebraic equations on a number plane. Some equations have graphs that are straight lines.

To graph an algebraic equation:

- construct a table of values for the equation
- plot the points from the table of values on a number plane
- rule a straight line through the points.

We can choose any numbers we want for x, but we need to make sure the points will fit on our graph and be easy to calculate. It is easiest to choose whole numbers close to 0.

EXAMPLE 1

Graph the equation $y = x + 2$.

Solution

Draw a table and choose some x values.

x	−2	−1	0	1	2	3
y						

Calculate the y values by substituting the x values in the equation.

$y = -2 + 2 = 0$ $\qquad$ $y = 1 + 2 = 3$

$y = -1 + 2 = 1$ $\qquad$ $y = 2 + 2 = 4$

$y = 0 + 2 = 2$ $\qquad$ $y = 3 + 2 = 5$

Complete the table.

x	−2	−1	0	1	2	3
y	0	1	2	3	4	5

Write the points from this table.

$(-2, 0), (-1, 1), (0, 2), (1, 3), (2, 4), (3, 5)$

Draw a set of axes and plot the points. Rule a straight line through the points, place arrows at each end and label the line with its equation.

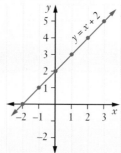

NELSON QMATHS 11. Essential Mathematics $\qquad$ ISBN 9780170412650

Gradient and *y*-intercept

The **gradient** of a line measures how steeply the line goes up or down.

The *y*-**intercept** of a line is the value where the line crosses the *y*-axis.

Gradient and
y-intercept
of a line

A **positive gradient** means the line goes up from left to right.

A **negative gradient** means the line goes down from left to right.

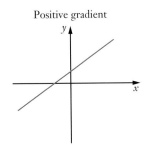
Positive gradient

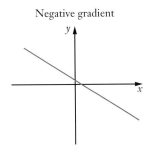
Negative gradient

EXAMPLE 2

This line has the equation $y = 2x + 1$.

a What is the gradient of the line?

b What is the value of the *y*-intercept?

c How do your answers to **a** and **b** relate to the equation of the line?

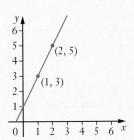

Graph
gradients

WS

Drawing
gradients

Solution

a Draw a right-angled triangle off the 2 points using the line as the hypotenuse.

The line goes up 2 units for every 1 unit across to the right.

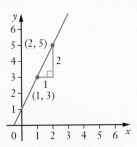

$$\text{Gradient} = \frac{2}{1} = 2$$

The gradient is 2.

b The *y*-intercept is the value where the line crosses the *y*-axis.

The *y*-intercept is 1.

ISBN 9780170412650

c The equation of the line is $y = 2x + 1$.

The gradient 2 is the number in front of the x in the equation.

The y-intercept 1 is the number on its own in the equation.

Gradient goat

WS

Gradient and y-intercept

EXAMPLE 3

Find the gradient and y-intercept of the line with equation:

a $y = 4x - 1$ **b** $y = -2x$ **c** $y = 8 - 3x$

Solution

a The gradient is the number in front of the x.

The y-intercept is the **constant** or the number on its own.

$y = 4x - 1$

gradient $= 4$

y-intercept $= -1$

b

$y = -2x$

gradient $= -2$

y-intercept $= 0$

c $8 - 3x$ can be rewritten as $-3x + 8$.

$y = -3x + 8$

gradient $= -3$

y-intercept $= 8$

Shutterstock.com/Mincemeat

Exercise 9.01 Straight-line graphs

1 Draw the graph of each equation.

   a $y = x - 4$ **b** $y = 2x$ **c** $y = x + 3$

   d $y = \dfrac{x}{2}$ **e** $y = -3x$ **f** $y = 2x - 4$

   g $y = -x + 2$ **h** $y = 3x$ **i** $y = \dfrac{x}{4}$

2 For each line, find its gradient and y-intercept.

   a

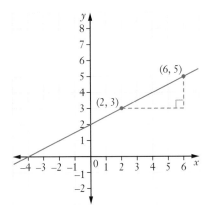

   b

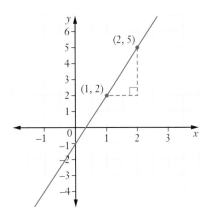

   c

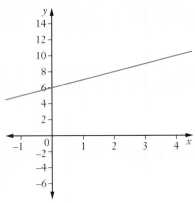

   d

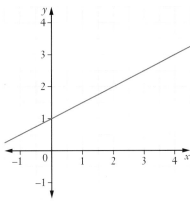

3 For each equation, state:

   i   the gradient   **ii**   the y-intercept.

   a $y = 3x + 2$ **b** $y = -2x + 3$ **c** $y = 7 - x$

   d $y = 4x$ **e** $y = \dfrac{x}{4} - 2$ **f** $y = -\dfrac{x}{3} - 4$

> Remember: $\dfrac{x}{4} = \dfrac{1}{4}x$

4 For each graph you drew in question **1**, state whether the gradient is positive or negative and write the gradient from the equation.

Linear
modelling

9.02 Using straight-line graphs

Often, straight-line graphs can be used to model real-world situations.

EXAMPLE 4

The concentration of a particular drug in a person's body decreases as time passes. This is represented by the graph shown, which has equation $C = -25t + 200$. Note that the variables x and y have been replaced by t and C respectively.

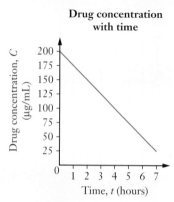

Drug concentration with time

a   What is the gradient of the line?

b   What does the gradient represent?

c   Find the **vertical intercept** of the line.

d   What does the vertical intercept represent?

e   When will there be no drug remaining in the body?

μg means micrograms or one millionth of a gram.

Solution

a   Draw a triangle on the line to calculate the gradient.

The line goes down 125 units for every 5 units across. The gradient will be negative because the line goes down rather than up.

OR

Read the gradient from the equation of the line. It is the number in front of t.

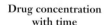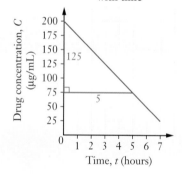

Drug concentration with time

$$\text{Gradient} = -\frac{125}{5}$$
$$= -25$$

b

The gradient represents the decrease in drug concentration per hour. It goes down 25 μg/mL every hour.

ISBN 9780170412650

c The vertical intercept is the general name for the y-intercept. It is the value where the line crosses the vertical axis and is the constant in $C = -25t + 200$.

The vertical intercept is 200.

d The vertical intercept is the drug concentration when $t = 0$.

The vertical intercept represents the initial concentration of the drug: 200 µg/mL.

e There is no drug remaining when $C = 0$. We can either extend the line to see where it crosses the horizontal axis, or substitute $C = 0$ into the equation and solve the equation.

When $C = 0$,

$$0 = -25t + 200$$
$$25t = 200$$
$$\frac{25t}{25} = \frac{200}{25}$$
$$t = 8.$$

There will be no drug in the body after 8 hours.

EXAMPLE 5

Carly has a window cleaning business. Each day it costs her an average of $75 for fuel and car maintenance and $8 per job.

Getty Images/Veresovich

a Let C = Carly's daily costs in dollars and n = the number of jobs she does per day.
Write an equation for Carly's daily costs.

b Construct a graph to illustrate Carly's daily costs.

c What is the gradient and vertical intercept of the graph and what do these values represent?

---

9. Practical graphs

Solution

a Carly's daily costs
= \$75 car costs + \$8 per job

$C = 75 + 8 \times n$
$C = 8n + 75$

b Complete a table of values for
$C = 8n + 75$.

n	0	1	2	3	4
C	75	83	91	99	107

Use the table of values to
graph a line on a number
plane.

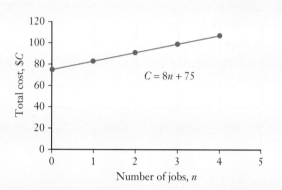

c For $C = 8n + 75$, the gradient is 8 and the vertical
intercept is 75.

The gradient represents Carly's costs per job (\$8).
The vertical intercept represents Carly's fixed
daily costs (\$75); the cost without doing any jobs.

Exercise 9.02 Using straight-line graphs

Example 4

1 The number of times a cricket chirps per minute
is related to the temperature. The relationship is
represented by the equation $n = 8T + 40$, where T is the
temperature in °C and n is the number of chirps. This
is shown on the graph at right.

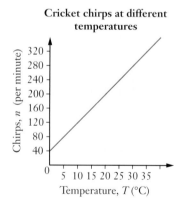

Cricket chirps at different temperatures

a What is the graph's vertical intercept?

b What does the vertical intercept represent in this
situation?

c Calculate the gradient of the line.

d In this context, what does the gradient represent?

e How many times per minute does a cricket chirp
when it is 32°C?

f At what temperature do crickets chirp 160 times per minute?

g Using the formula, calculate how many times per minute a cricket chirps at 100°C.
Is this realistic?

2 At sea level, water boils at 100°C. At different altitudes, the boiling temperature changes.

| 'Altitude' is another word for 'height'. |

This graph shows the relationship between altitude (h km) and the temperature (T°C) at which water boils. The equation of the line is $T = -3.5h + 100$.

a Water boils at 100°C at sea level. What altitude is sea level?

b Calculate the gradient of the line.

c What physical quantity does the gradient represent?

d Use the graph to determine the boiling point of water at an altitude of 2.8 km.

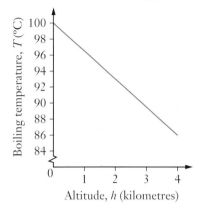

Boiling point of water at different altitudes

3 After school, 3 students went home by walking, jogging or riding a bike. Each of them travelled at a constant speed and took one hour to reach home. The graph shows the distance they travelled.

a Calculate the gradient of each line.

b In this context, what does the gradient represent?

c Which student travelled at the slowest speed? Give a reason for your answer.

d Match Laura, Bevan and Jacob to their method of travelling home: walking, jogging or riding a bike.

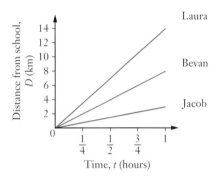

4 This graph shows the length and head circumference of average newborn baby boys.

Newborn baby Jason is 51 cm long and his head circumference is 35.1 cm. Do you think Jason's mother should be worried about him? Give a reason for your answer.

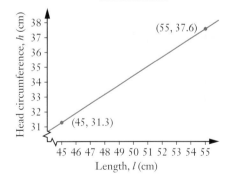

Baby boys' length and head circumference

5 Anastasia operates a printing business. When she quotes a price for printing tickets, she charges an initial fee to cover the cost of design and an additional fee of $0.25 per ticket printed. The price she charges for printing tickets is shown on the graph.

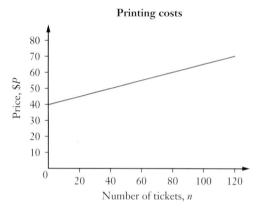

Printing costs

a How much is Anastasia's initial charge for the design?

b What is the gradient of the line?

c What physical quantity does the gradient represent?

d How much does she charge to design and print 500 tickets?

e When Anastasia designs and prints elaborate wedding invitations, she charges an initial fee of $110 and 50c per invitation. Write a formula to determine the price, in D dollars, for designing and printing n elaborate wedding invitations.

6 Alan runs a coffee stall at sporting events. The graph shows how much it costs Alan to set up the stall and make mugs of coffee.

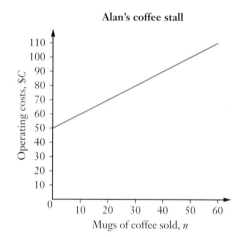

Alan's coffee stall

a How much does it cost Alan to set up the stall?

b How much does it cost Alan to make each mug of coffee?

c If n is the number of mugs sold and C are Alan's costs in dollars, what is the equation of the line on the graph?

7 Belinda's hobby is jumping out of planes! This table of values shows her height above the ground t seconds after she opened her parachute.

Example 5

Time t seconds	0	60	90	120
Height, h metres	600	300	150	0

a How high above the ground was Belinda when she opened her parachute?

b How many metres is Belinda falling every second?

c Explain why the equation $h = 600 - 5t$ represents this situation.

d Graph this equation.

Shutterstock.com/Germanskydiver

8 Gabrielle is investing $2000 at a simple interest rate of 3.5% p.a. This table shows the interest earned using the formula $I = Pin$.

Number of years (n)	0	1	2	3	4	5
Interest earned (I)	0	$70	$140	$210	$280	$350

a Graph this function.

b Find the gradient of the line and explain what this value represents.

c Find the vertical intercept and explain what this value represents.

d Use the equation to find the interest earned if $2000 is invested at 3.5% p.a. for 7 years.

e Gabrielle withdraws her money after 2 years. How much money in total does she receive?

9 The cost of filling your petrol tank is related to the number of litres it holds, as shown in this table.

Petrol amount (*L* litres)	0	10	20	30	40	50
Cost ($*C*)	0	14	28	42	56	70

a Graph the relationship between petrol amount and cost.

b What is the gradient of the graph?

c Calculate the cost of buying 75 litres of petrol.

10 Amy has old dressmaking patterns that show material measurements in yards, an imperial measure, which she needs to convert to metres.

a Use this table of values to construct a conversion graph between yards and metres.

Yards	0	1	3	5	10
Metres	0	0.9	2.7	4.5	9.1

b Use your graph to estimate the number of metres of material Amy needs when the pattern states $2\frac{3}{4}$ yards.

c Amy has 4.2 m of material left for making a dress. The pattern says she needs 4 yards of material. Does Amy have enough material to make the dress? Justify your answer.

9.03 Intersecting graphs

We can graph 2 lines on the same number plane to help us solve problems.

Intersection of lines

EXAMPLE 6

Cherie sells scones for $3 each. It costs her $240 for the necessary equipment and $1 to make each scone.

a Find an equation for C, the cost in dollars for making n scones.

b Find an equation for I, the income in dollars from selling n scones.

c Graph both equations on the same axes for values of n from 0 to 200.

d How many scones does Cherie need to sell to break even?

e How much profit will Cherie make if she sells 390 scones?

Solution

a The cost is $240 plus $1 for each scone Cherie makes.

$$\text{Cost} = \$240 + n \times \$1$$
$$C = 240 + n$$
$$C = n + 240$$

b Income is $3 for each scone.

$$\text{Income} = n \times \$3$$
$$I = 3n$$

c Graph both lines together.

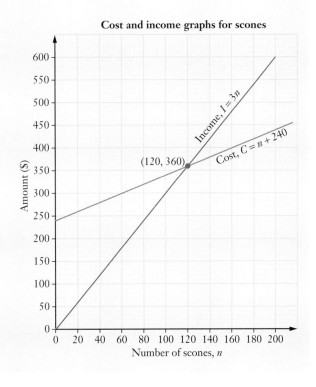

Cost and income graphs for scones

d 'Break even' means that costs and income are the same. This is the point where the lines intersect, at (120, 360), where 120 represents the number of scones.

Cherie needs to sell 120 scones to break even.

e The profit is how much the income is more than the cost: substitute $n = 390$ into both the Income and Cost equations.

Income: $I = 3n$
$= 3 \times 390$
$= \$1170.$
For 390 scones, the income will be $1170.
Cost: $C = n + 240$
$= 390 + 240$
$= 630$
For making 390 scones, the cost will be $630.

Profit = income − cost

Profit = $1170 − $630
$= \$540$
Cherie will make a $540 profit when she makes and sells 390 scones.

Break-even points

In the above example, when $n = 120$, $C = 360$ and $I = 360$. The cost and income are the same. This means that when Cherie makes 120 scones, it costs her $360 and she receives $360 from the sales. From the graphs, we can see that if she sells more than 120 scones, she will make a profit because her income is greater than her costs, but if she sells fewer than 120 scones, she will make a loss because her income is less than her costs. The value of 120 is called the **break-even point**.

If we graph the equations of a business' costs and income, then the break-even point is the point where the 2 lines intersect. Below the break-even point, the costs are greater than the income and the business makes a loss. Above the break-even point, the income is greater than the costs and the business makes a profit.

Exercise 9.03 Intersecting graphs

Example 6

1 Khalid bakes and sells muffins. The equipment used to bake the muffins costs $200 and each muffin he makes costs $1. Khalid sells the muffins for $2 each.

Let n = the number of muffins, $\$C$ = the total cost to make n muffins and $\$I$ = the income from selling n muffins.

a Explain why $C = n + 200$ and $I = 2n$.

b Graph $C = n + 200$ and $I = 2n$ on the same axes using n from 0 to 400.

c What are the coordinates of the point where the 2 lines intersect?

d What does this point represent?

e Will Khalid make a profit if he sells 120 muffins? Explain your answer.

f How much profit will Khalid make if he sells 230 muffins?

2 Chloe is deciding between 2 mobile phone plans. Plan *A* offers her no connection fee and charges of 40c per minute for each call. Calls under plan *B* cost 20c per minute with a 60c connection fee. The graphs show the costs of phone calls under each plan for calls of up to 6 minutes.

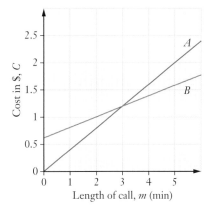

a Which plan (*A* or *B*) has the equation $C = 0.6 + 0.2m$? Explain the reason for your choice.

b For what length of phone call is the cost the same under both plans?

c Which plan is cheaper for longer calls?

3 A truck and a car were travelling on the freeway. When an accident occurred ahead of them, both vehicles had to stop quickly. The graphs show the speed each vehicle was travelling at *t* seconds after they applied their brakes.

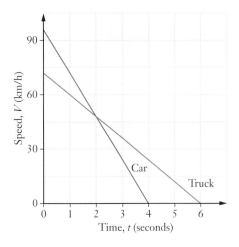

a How fast were the car and the truck going in km/h before the accident?

b How many seconds after applying their brakes were the vehicles travelling at the same speed?

c How much longer than the car did it take for the truck to stop?

d Write a linear equation for V, the speed in km/h, for the speeds of each vehicle t seconds after applying their brakes.

e Explain why the gradients of both lines are negative.

f Suggest a reason why the truck took longer to stop than the car, even though the car was travelling faster initially.

4 Allana sells flowers at the market for $10 per bunch.

a Copy and complete this table.

Bunches of flowers sold	0	10	20	30	40	50	60
Income received ($)							

b This graph shows Allana's *costs* when she sells n bunches of flowers. Copy the graph and use your answers to part **a** to graph Allana's income on the same axes.

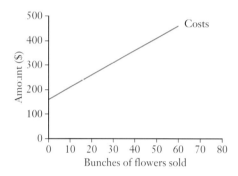

c Explain how you know that Allana will break even when she sells 32 bunches of flowers.

d How many bunches of flowers will Allana need to sell to make a profit?

5 Each week, a toy factory's fixed costs total $1500. Its costs increase by $15 for every toy it produces. Each item produced sells for $35.

Let n = the number of items the factory produces each week, C = the total costs to produce n items, and I = the income the factory receives from selling n items.

a Does the expression $35n$ represent the income from selling n items or the cost to produce n items?

b What does the expression $15n + 1500$ represent?

c Graph both equations $I = 35n$ and $C = 15n + 1500$ on the same axes, using n from 0 to 200.

d How many toys does the factory need to produce each week to break even?

e How much profit will the company make if it sells 100 toys in a week?

6 Jon earns money taking tourists on dog sled rides in the snow. Jon has displayed his expenses and income on a graph.

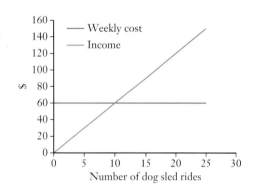

Number of dog sled rides	0	5	10	15	20	25
Weekly cost	60	60	60	60	60	60
Income	0	30	60	90	120	150

a How much are Jon's weekly costs?

b How many dog sled rides does Jon require each week to break even?

c How much does Jon charge for a dog sled ride?

d Suggest a reason why Jon's cost line is horizontal.

7 To advertise the school musical, the drama class is making and selling promotional T-shirts. The set-up costs to make the T-shirts is $160 and each shirt will cost $10 to make. The class will sell the T-shirts for $20 each.

a Find an equation for C, the cost in dollars for making n T-shirts.

b Find an equation for I, the income in dollars from selling n T-shirts.

c Graph both equations on the same axes, for values of n from 0 to 50.

d How many T-shirts need to be sold to break even?

e The class estimates it will sell 90 T-shirts. How much profit can they expect to make?

8 Grant is hard of hearing and uses his mobile phone for text messages only. This graph shows the monthly cost, D dollars for n text messages on 2 phone plans. Plan A is a fixed price per month with unlimited texts while Plan B has a monthly charge and a cost per text.

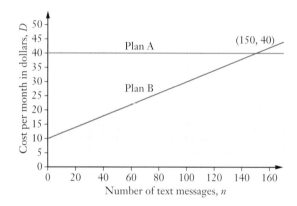

a What is the monthly cost for Plan A?

b What is the vertical intercept of the graph of Plan B?

c What does the vertical intercept in Plan B represent?

d How much does each message cost to send on Plan B?

e What is the equation of the line representing Plan B?

f For what number of calls per month is Plan A the same cost as Plan B?

g Grant sends an average of 5 or 6 texts per day. Which plan would you recommend he chooses? Use calculations to justify your recommendation.

Chapter problem

You've covered the skills required to solve the chapter problem. Can you solve it now?

COSTING THE YEAR 12 FORMAL

In this investigation you will model the costs and income for running the Year 12 formal.

1 You will need to collect the following information for where you live:

Costs   Hire of venue   _____
                 Decorations   _____
                 Band/DJ   _____
                 Other costs   _____
                 Total costs   _____

These are the fixed costs for the formal.

The committee decides to charge $15 per person to cover the fixed costs.

2 On the same number plane:

- draw a graph for $I = 15n$ for n from 0 to 200.

- draw a graph for $F =$ _____ for the fixed costs. This will be a horizontal line.

3 a Where do the 2 lines intersect? How many people need to attend the formal to cover the fixed costs?

b How many people do you expect to attend the formal? Will you cover the fixed costs if you charge $15 per person?

c Depending on what you have found, adjust the amount you need to charge per person to cover the fixed costs. This could be more or less than $15. Draw the income graph for this new amount.

d How many people now need to attend the formal to cover the fixed costs?

4 Find out from the venue how much they charge per person for catering and calculate what you need to charge for your formal tickets.

Shutterstock.com/sirtravelalot

9.04 Curved graphs

Not all practical situations give straight-line graphs. Now we will look at some situations that when graphed give us curves rather than straight lines.

Quadratic functions

EXAMPLE 7

A rock fell from the top of a cliff onto a road. The graph shows the height of the rock above the road t seconds after it fell.

a How high was the rock above the road after 2 seconds?

b How far did the rock fall during those first 2 seconds?

c Approximately after how many seconds did the rock hit the road?

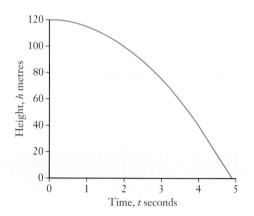

Solution

a Read from the graph.

The rock was 100 m above the road after 2 seconds.

b The rock was 120 m high before it fell and 100 m high after 2 seconds.

During the first 2 seconds it fell:
$120 \text{ m} - 100 \text{ m} = 20 \text{ m}$

c Read from the graph.

The rock hit the road at about 4.9 s.

EXAMPLE 8

Isobel plays basketball. When she takes a shot, the ball follows a curve. This table gives the height, h metres, of the ball t seconds after it leaves her hands.

Time, t (seconds)	0	0.5	1	1.5	2	2.5
Height, h (metres)	2.4	3.3	3.7	4.0	3.6	2.8

Graph this data and join the points with a smooth curve.

Solution

Draw a set of axes with Time on the horizontal axis and Height on the vertical axis.

Plot the points.

Join them with a smooth curve.

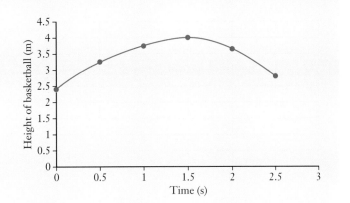

Exercise 9.04 Curved graphs

1 When Jose jumps out of a plane, he free-falls for 15 seconds. The graph shows his height during the free-fall.

a At what height did Jose jump out?

b How high was Jose above the ground 10 seconds after he jumped?

c How far did Jose fall in the first 10 seconds?

d Approximately how high above the ground was Jose when he pulled his parachute chord?

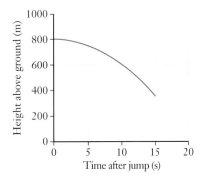

Example
7

2 Adriana is planning to make a square chicken shed on her farm. The graph shows the relationship between the length and area of the shed.

Use the graph to estimate the length of the chicken shed with an area of 30 m².

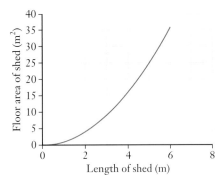

3 This graph represents part of a roller coaster track at a theme park.

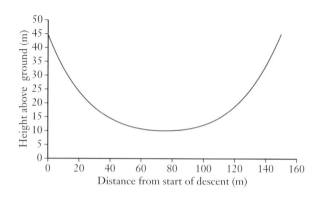

a At what height does this section of the ride start?

b How high above the ground is the ride when it has travelled 50 m horizontally?

c How far has the ride travelled horizontally when it reaches this height again?

4 Naresh hit a cricket ball out of the oval. The graph shows the path of the ball.

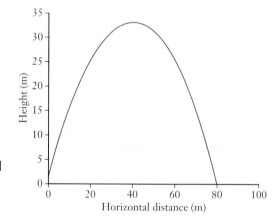

a Estimate the maximum height of the ball.

b Approximately how far from Naresh did the ball land?

c While falling, the ball just cleared the top of the scoreboard, which is 12 m high. How far is the scoreboard (horizontally) from Naresh?

5 This table shows the height of tides in Cairns on 3 consecutive days.

Day	Wednesday				Thursday				Friday			
Time	1 a.m.	7:30 a.m.	1:15 p.m.	7:45 p.m.	1:45 a.m.	8 a.m.	2 p.m.	8:15 p.m.	2:15 a.m.	8:45 a.m.	2:30 p.m.	8:45 p.m.
Height of tide (m)	0.5	2.6	0.6	2.8	0.4	2.7	0.6	2.7	0.4	2.8	0.8	2.6

a Graph this data with Time on the horizontal axis and Height on the vertical axis. Join the points with a smooth curve.

b Use your graph to estimate the height of the tide at midnight Wednesday.

c Joe needs a tide height of 1 m to go fishing. Use your graph to estimate when he can go fishing on Friday afternoon.

6 For a Science project, Ellie measures the temperature of a cup of coffee at different times after she makes it. Her data is given in the table below.

Time (min)	0	4	8	12	16
Temperature (°C)	80	52	47	43	40

a When you graph this table, should Time or Temperature be on the horizontal axis?

b Graph this table and join the points with a smooth curve.

c Use your graph to estimate the temperature of the coffee after 2 minutes.

d Ellie likes to drink her coffee at approximately 53°C. After she makes her coffee how long should she wait to drink it?

7 These tables show the hours and minutes of daylight in 2 cities on the 21st day of the month over one year. For example, 6:52 means 6 hours 52 minutes of daylight.

Anchorage, Alaska											
Jan	Feb	Mar	Apr	May	Jun	Jul	Aug	Sep	Oct	Nov	Dec
6:52	9:39	12:20	15:18	17:59	19:22	18:00	15:16	12:21	9:31	6:47	5:27

Perth, Australia											
Jan	Feb	Mar	Apr	May	Jun	Jul	Aug	Sep	Oct	Nov	Dec
14:39	13:25	12:07	10:43	9:37	9:09	9:36	10:42	12:05	13:27	14:40	15:11

a On the same set of axes, graph the data for each city and join the points with a smooth curve. Use a different colour for each city.

b What are the similarities between the 2 graphs?

c What are the differences between the 2 graphs?

d In which months do the graphs cross?

e Why is the maximum hours of daylight in Anchorage in June greater than the maximum hours of daylight in Perth in December?

8 Toby is organising a bus trip to see a concert. The cost of the bus is $840 and the maximum number of people it carries is 40.

a Copy and complete this table to show the cost per person for different numbers of people.

Cost per person = 840 ÷ the number of people

Number of people	1	5	10	20	30	40
Cost per person ($)	840	168				

b Graph this data, joining the points with a smooth curve.

c From your graph estimate the cost per person if 28 people attend.

d Calculate the cost using the formula. How close was your estimate from the graph?

9.05 Practical graphs

Some situations give graphs that don't follow an equation. However, we can draw a graph from a description of what is happening or we can work out the story of what is happening when we see the graph.

EXAMPLE 9

Sarah is pouring water into this container at a constant rate. Draw a graph showing the height of the water in the container over time.

Solution

At first, the height of the water will increase at a steady rate.

When the water reaches the narrower section of the container, the height will go up more quickly, at a constant faster rate.

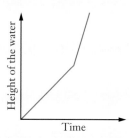

NELSON QMATHS 11. Essential Mathematics

ISBN 9780170412650

EXAMPLE 10

This graph shows the amount of petrol in the tank of Natasha's car over a week, starting on Monday. Describe what the graph is showing.

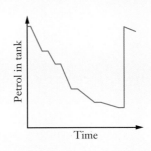

Solution

The amount of petrol goes down over a number of days. It then increases quickly before decreasing a small amount.

Other descriptions are possible.

Natasha starts the week with a full petrol tank. As she drives over the week, the amount of petrol in the tank goes down.

When the graph is horizontal (flat), Natasha isn't using the car.

Towards the end of the week, Natasha fills the tank and drives again.

Exercise 9.05 Practical graphs

1 Water is poured into each container at a steady rate. Match the container to the graph of the height of the water in the container.

Example 9

a

b

c

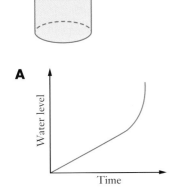
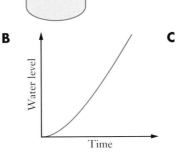
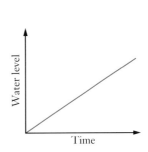

A

B

C

2 Filomena poured orange juice into this jug before her party. Draw a graph showing the height of the orange juice in the jug over time if she pours the orange juice at a constant rate.

3 A grain silo at a train siding is emptied at a constant rate into a goods train to be transported to the city. Draw a graph showing the height of the grain in the silo over time.

Example **10**

4 This graph shows the speed of a train as it travels from Oxford Park to Ferny Grove. Write a description of what happens.

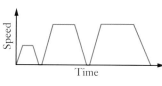

PS

5 This graph shows the number of students at Nelson Valley High over 1 day. Write a description of how the number of students changes over the day.

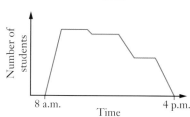

6 This graph shows the temperature in Canberra over one day in July from midnight. Write a story to describe the change in temperature based on this graph.

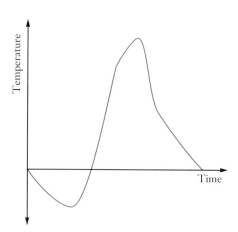

PS

7 Draw a graph similar to that in Question **6** to show the change in temperature in your town or city over one day.

8 This incomplete graph shows Hong's trip from Hervey Bay to Maroochydore and back.

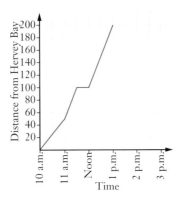

a Describe the section of the journey shown on the graph.

b Hong stays in Maroochydore for 2 hours and then returns to Hervey Bay without stopping, arriving at 5 p.m. Copy the graph, extending the Time axis to 5 p.m., and complete the graph to show the return section of Hong's trip.

9 A lift is going up an 8-storey building. On ground level, 5 people enter the lift. 2 people get off on Level 2. One person gets off at Level 5. Two people get off on Level 8 where 3 other people get on the lift. They travel back to the ground level.

Construct a graph to represent the height of the lift. Use Time on the horizontal axis and Building level on the vertical axis.

KEYWORD ACTIVITY

CHAPTER SUMMARY

Use the listed words to copy and complete the summary of this chapter below.

algebraic	break-even	costs	curved
expenses	gradient	income	intersecting
loss	model	point	profit
straight	y-intercept		

In this chapter, we learned about [1]_____ equations that give us [2]_____ lines when we graph them. The [3]_____ of the line tells us how steep the line is. The [4]_____ tells us where the line crosses the y-axis.

We used straight-line graphs to [5]_____ real-life contexts.

For two [6]_____ lines, we examined the [7]_____ where these lines crossed. We applied this idea to practical problems such as business, travelling and comparing mobile phone plans.

In business, we drew a graph for the money coming in, called [8]_____, and the money being spent, called [9]_____ or [10]_____. The point of intersection of these two lines is called the [11]_____ point. At this point, the business owner does not make a [12]_____ or a [13]____.

We also learned about practical situations that don't have a straight-line graph. Instead, many graphs are [14]_____.

SOLUTION TO THE CHAPTER PROBLEM

Problem

Paul owns a factory that makes sails. It costs Paul $9000 per month to cover rent, electricity and wages. Each sail costs $250 to make and sells for $850.

a   How many sails does Paul need to make and sell per month to break even?

b   What profit or loss will he make in a month when he makes and sells 20 sails?

Solution

STAGE 1: WHAT IS THE PROBLEM? WHAT DO WE KNOW?

a   We need to find the break-even point. To do this, we draw the graphs for cost and income.

WHAT?   b   We need to calculate the profit or loss if 20 sails are sold.

STAGE 2: SOLVE THE PROBLEM

Let n = the number of sails, C = the cost to produce n sails per month, and I = the income from selling n sails.

SOLVE   a   The linear functions are:

$C = 250n + 9000$

$I = 850n$

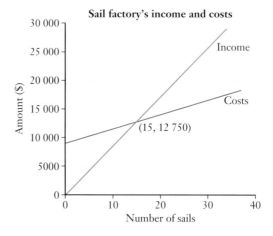

The lines intersect at (15, 12 750), which means that the break-even point is $n = 15$.

Paul needs to make 15 sales per month to break even.

b When $n = 20$:

Cost $= \$250 \times 20 + \$9000$

$= \$14\ 000$

Income $= \$850 \times 20$

$= \$17\ 000$

Income is greater than cost, so a profit will be made.

Profit $= \$17\ 000 - \$14\ 000$

$= \$3000$

The factory will make a profit of $3000 when it makes and sells 20 sails in a month.

STAGE 3: CHECK THE SOLUTION

Income is above costs on the graph when $n = 20$, so solution is correct.

We have answered both parts of the question.

CHECK

STAGE 4: PRESENT THE SOLUTION

a Paul needs to sell 15 sails per month to break even.

b The factory will make a profit of $3000 when it makes and sells 20 sails in a month.

PRESENT

Practical graphs

1 Draw the graph of:

a $y = 2x - 2$ **b** $y = -x + 3$

2 For each graph find:

i the gradient   **ii** the y-intercept

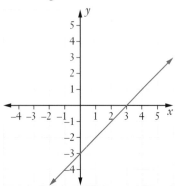 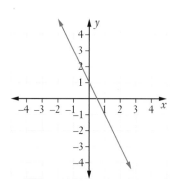

3 For the graph of each equation, state:

i the gradient   **ii** the y-intercept

a $y = 4x - 3$ **b** $y = -x + 2$ **c** $y = \dfrac{x}{3}$

4 Goran packs glass items in boxes. He has to assemble the box before he can pack it. The graph shows the time, T minutes, it takes him to assemble one box and pack n glass items in it.

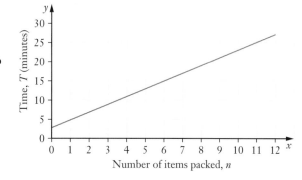

a How long does it take Goran to assemble a box before he starts to pack it?

b Calculate the gradient of the line.

c What physical amount does the gradient represent?

d Use the graph to find how many items can be packed in 15 minutes.

e Calculate the time it takes Graham to assemble a box and pack 15 items in it.

5 Adrian's Awesome Appetisers charges $10 per person plus an $80 fixed charge. This table shows Adrian's charges.

Exercise 9.02

Number of people, n	0	50	100	150	200	250
Charge, $C	80	580	1080	1580	2080	2580

a Construct a graph showing Adrian's charges for n people.

b Use the graph to find the cost for 170 people.

c Adrian charged NQM Bank $830 for catering for a party. How many people were catered for?

d Find the vertical intercept of the line and explain what this value represents.

e Find the gradient and explain what this value represents.

6 The Year 12 fundraising committee plans to sell school souvenir USB drives. The manufacturer quoted an 'initial charge of $126 plus $5 per unit' and the committee thinks that it will sell the units for $12 each.

Exercise 9.03

a Find an equation for $C, the cost of making n USB drives.

b Find an equation for $I, the income from selling n USB drives.

c Graph both equations on the same axes for values of n from 0 to 20.

d How many USB drives does the committee need to sell to break even?

e There are 140 students in Year 12. If half of them buy a USB drive each, how much profit can the committee expect to make?

7 This graph shows the braking distance for a car at different speeds.

Exercise 9.04

a What is the braking distance at 100 km/h?

b At what speed is the car travelling if the braking distance is 30 m?

c Bridie is travelling at 60 km/h. She sees an accident ahead. She travels 25 m before she applies the brakes. What is her total stopping distance?

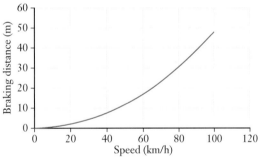

8 Susie planted a new shrub in her garden. She measured its height each week.

Exercise 9.04

Week	1	2	3	4	5	6
Height (cm)	3	7	10.5	13.5	16	18

Draw a graph of this data and join the points with a smooth curve.

9 Draw a graph that represents this story.

The noise level in a classroom varies over the one hour of a lesson. The class is quite noisy as they enter the classroom. They then become silent to listen to the teacher explain what is going to happen in the lesson. Some students ask questions before the class is divided into groups to work together for the rest of the lesson. Group work is very noisy. 5 minutes before the end of the lesson, the teacher asks for silence so that she can finish the lesson.

10 This shows the journeys of 2 cyclists on a Sunday bike ride. Describe the journeys of both cyclists.

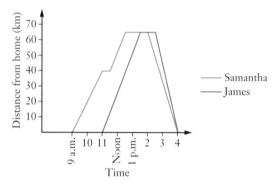

Practice set 2 ●●○○

Section A Multiple-choice questions

For each question select the correct answer **A**, **B**, **C** or **D**.

1 Which ratio below is NOT equivalent to 32 : 48?

   A 16 : 24   **B** 4 : 6   **C** 2 : 3   **D** 6 : 8

Exercise 6.01

2 Calculate the simple interest earned on $600 at 5% p.a. for 4 years.

   A $20   **B** $30   **C** $120   **D** $3000

Exercise 7.03

3 Which of the following is NOT numerical data?

   A The star ratings given by the audience of a movie from ★ to ★★★★★

   B The average temperature for each month

   C Rainfall data for Rockhampton measured in millimetres

   D The number of shoes each student owns

Exercise 8.01

4 Calculate $12\frac{1}{2}\%$ of 6 m.

   A 0.75 mm   **B** 7.5 mm   **C** 75 mm   **D** 750 mm

Exercise 7.01

5 A cup of tea sits on the kitchen bench cooling. At first it loses heat quickly, but as time passes, it loses heat more slowly until it reaches room temperature. Which graph best illustrates this?

Exercise 9.05

   A

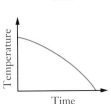

   B

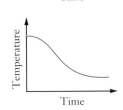

   C

   D

6 Alison and Elena buy a length of material and divide it between themselves in the ratio 2 : 3. Alison has 3.6 m. What length of material does Elena have?

   A 9 m   **B** 5.4 m   **C** 2.4 m   **D** 1.8 m

Exercise 6.03

7 This graph shows the income and expenditure for Catriona's Cupcakes.

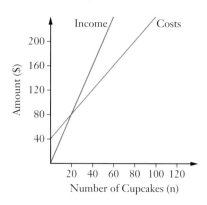

What happens if Catriona sells 40 cupcakes?

A She makes a profit   **B** She makes a loss

C She breaks even   **D** You can't tell from the graph

8 A street map has a scale of 1 : 200 000. What distance on the map represents 9.6 km?

A 0.048 cm   **B** 0.48 cm   **C** 4.8 cm   **D** 48 cm

Section B Short-answer questions

1 Marika is saving for a holiday. So far, she has saved $3500. She invests this money for 6 months at a simple interest rate of 2% p.a.

  a How much interest will she earn?

  b How much will she have in total after 6 months?

2 Identify the outlier(s) in this data set showing the number of goals scored in each match by a junior soccer team.

2	1	0	0	7	2	3	2	4
3	2	9	1	2	1	2	4	3

3 A catapult is a medieval weapon for hurling large stones. When the stone travels in the air, it follows a curve. This table gives the height, h metres, of the stone d metres horizontally from the catapult.

Horizontal distance from the catapult (*d* metres)	0	1	2	3	4	5
Height (*h* metres)	5	6.5	7	6.5	5	2.5

  a Draw the graph of this data and join the points with a smooth curve.

  b What is the greatest height reached by the stone?

4 For their new bathroom, Ann and Chris need 4 green tiles for every 3 white tiles. They buy 100 green tiles. How many white tiles do they need to buy?

5 Sam rolled a die 20 times and recorded the uppermost number. These are the results:

$$5, 2, 3, 4, 2, 2, 6, 2, 6, 3, 5, 1, 4, 1, 2, 5, 6, 2, 4, 1$$

a Draw a dot plot for this data.

b How many times did Sam roll a number greater than 3?

c In what percentage of rolls did Sam roll a 1 or a 2?

6 Franken's Furniture buys a lounge suite from a supplier for $2300. The store adds a 60% mark-up to obtain the selling price. Then 10% of the selling price must be added for GST.

a Calculate the selling price of the lounge suite before GST is added.

b Calculate the price including GST.

c In a stocktake sale in June, the lounge suite is discounted by 30%. Find its discounted selling price.

7 Year 11 is running a sausage sizzle at the athletics carnival to raise money for charity. It will cost them $32 for the gas bottle for the barbecue and each sausage in bread will cost $0.90 to make. This table shows the costs for making up to 200 sausages in bread.

Number of sausages in bread, n	0	10	20	30	40	50	100	150	200
Cost, $C	32	41	50	59	68	77	122	199	244

a Graph this table of values.

b Use the graph to estimate the cost of making 170 sausages in bread.

c Find the vertical intercept and explain what this value represents.

d Find the gradient and explain what this value represents.

e Year 11 sold 75 sausages in bread for $2 each at the carnival. Did they make a profit or a loss? Justify your answer.

8 Jemma likes to mix a paint colour called Spring Green. She mixes blue and yellow paint in the ratio 5 : 3. How much blue paint does Jemma use to make a 4 litre can of Spring Green? Give your answer in millilitres.

9 In order to improve customer service, the NQM Bank recorded the waiting times, in minutes, per customer at one of its biggest branches.

14 11 12   7   6 13 12 21 11   6 13 11   7   3   8 11 10   8   9   7   9
11 10   5 14 12 12 10 13 14   9 15   9 12   7 22   8   9 12   7   6   9

a Copy and complete this frequency table for the data.

Waiting time	Tally	Frequency
1–5		
6–10		
11–15		
16–20		
21–25		

b For how many customers was the waiting time from 11 to 15 minutes?

c What percentage of customers waited for over 20 minutes? Answer correct to one decimal place.

d Draw a frequency histogram for this data.

10 This graph shows the height of water in a water tank over one month. Write a description of what the graph is showing.

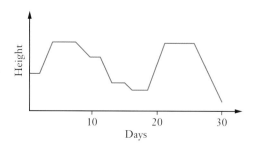

11 For an adult male, the ratio of the length of his head to his height is 1 : 8.
Police found an adult male skull in scrubland measuring 21.5 cm long. What was the approximate height of the man?

12 Rayna borrowed $26 000 from a finance company to start a small business. She was charged 5.2% p.a. simple interest and agreed to pay the loan back over 5 years.

Exercise
7.03

   a How much interest was Rayna charged by the company?

   b How much did she repay in total?

   c Rayna repaid the loan by making monthly payments. How much does she have to pay each month?

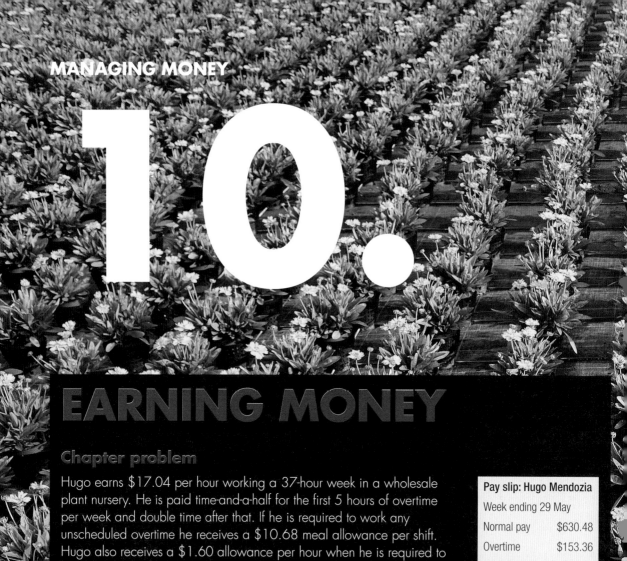

10.

EARNING MONEY

Chapter problem

Hugo earns $17.04 per hour working a 37-hour week in a wholesale plant nursery. He is paid time-and-a-half for the first 5 hours of overtime per week and double time after that. If he is required to work any unscheduled overtime he receives a $10.68 meal allowance per shift. Hugo also receives a $1.60 allowance per hour when he is required to work in wet areas.

In the week ending 29 May, he worked 43 hours, which included one unscheduled overtime shift and 6 hours working in a wet area.

Hugo thinks his pay slip for the week ending 29 May is wrong. Is his gross pay correct?

Pay slip: Hugo Mendozia	
Week ending 29 May	
Normal pay	$630.48
Overtime	$153.36
Allowances	$20.28
Gross pay	$804.12

WHAT WILL WE DO IN THIS CHAPTER?

- Use rates to calculate wages, salaries, overtime and piecework
- Apply percentages to calculate bonuses, allowances, commissions, royalties and holiday pay
- Calculate superannuation

HOW ARE WE EVER GOING TO USE THIS?

- Checking that pay is correct, including bonuses and holiday pay
- Comparing jobs and choosing the jobs with the best pay
- Contributing to superannuation as soon as we start work to provide for retirement

10.01 Wages and salaries

Wages and salaries are different ways we can be paid for work we do for an employer.

A **wage** is an amount paid for each hour worked.

A **salary** is a fixed amount per year that does not depend upon the number of hours worked. Teachers and other professionals earn a salary.

PROFILE

KRISTINE – CHILD CARE WORKER

When I was at school I earned money by babysitting. It was a job and it was fun. When I left school I decided to go into childcare. There are lots of jobs in childcare because most mums need to go to work. I did a Certificate III in Childcare at TAFE and now I have a job I love. The best part of my job is helping children learn.

EXAMPLE 1

Kristine earns $18.23 per hour working in a long-day care centre. She works a 38-hour week.

a How much does Kristine earn per fortnight?

b Calculate the amount Kristine earns per year.

Solution

a Calculate Kristine's earnings for 1 week.

Amount per week $= 38 \times \$18.23$
$= \$692.74$

A fortnight is 2 weeks.

Amount per fortnight $= 2 \times \$692.74$
$= \$1385.48$

Each fortnight, Kristine earns $1385.48.

b There are 52 weeks in a year.

Amount per year $= 52 \times \$692.74$
$= \$36\,022.48$

Each year, Kristine earns $36 022.48.

EXAMPLE 2

Madeleine is a social worker. Her annual salary is $59 000. Calculate:

a her monthly pay

b her fortnightly pay.

Solution

a There are 12 months in a year.

'Annual' means per year.

$$\text{Monthly pay} = \text{Annual salary} \div 12$$
$$= \$59\,000 \div 12$$
$$= \$4916.67$$

b There are 26 fortnights in a year.

$$\text{Fortnightly pay} = \text{Annual salary} \div 26$$
$$= \$59\,000 \div 26$$
$$= \$2269.23$$

1 year = 12 months

1 year = 52 weeks

1 year = 26 fortnights

Watch out! One month is not 4 weeks and one year is not 48 weeks. This is a common mistake and it's WRONG.

Alamy Stock Photo/Neil McAllister

PROFILE

PAUL – AN AUSTRALIAN VOLUNTEER ABROAD

I completed a building apprenticeship when I left school, but when I finished my trade I wasn't ready to settle down. I wanted to travel and see the world. I saw a news report about volunteers helping in developing countries and I decided to do my bit to help others. The aid agency paid my travel costs and provides my accommodation and the equipment I need. They also pay me a modest stipend. As a volunteer I'm making a difference in this community.

A *stipend* (pronounced 'sty-pend') is similar to a salary, but it is usually for a relatively small amount. People in religious orders and some volunteers receive a stipend.

Exercise 10.01 Wages and salaries

1 Scott is a qualified ambulance paramedic. He is paid $35 per hour for a 38-hour week.

   a How much does Scott earn per week?

   b How much is he paid per fortnight?

   c Calculate Scott's annual pay.

2 Suzanne is a solicitor. Her salary is $82 500 p.a.

   a How much does Suzanne earn per month?

   b Calculate her fortnightly pay.

   c How much does Suzanne earn per week?

3 Lance is paid a salary for being an office IT manager. Each week he earns $1300.

   a Calculate Lance's annual salary.

   b Explain why Lance's monthly pay is *not* $1300 × 4.

   c Divide Lance's annual salary by 12 to determine his monthly pay.

   d Lance's salary is based on 7 hours work per day, 5 days per week and 52 weeks per year. Calculate the pay rate per hour that is the basis of Lance's salary.

4 Zheng earns $15.61 per hour at the Chinese take-away.

   a Last week, Zheng worked 16 hours. How much did he earn?

   b Today, Zheng earned $70.25. How long did he work?

5 Ulla receives a yearly stipend of $22 860 from the university to assist her with her postgraduate study and research.

   a How much does Ulla receive per fortnight from the stipend?

   b The stipend isn't enough to cover all of Ulla's living expenses. She also works as a waitress for 4 hours per night, 2 nights per week. She earns $18.20 per hour as a waitress. Calculate Ulla's total fortnightly income.

6 The minimum wage for a beginning pest inspector is $595.70 for a 38-hour week. What is the minimum pay per hour for a beginning pest inspector?

Shutterstock.com/Andrey_Popov

7 Carlos earns $320 per day as a relief teacher. The table shows the number of days he worked over 5 weeks.

Dates	Number of days worked
April 30 – May 4	2
May 7 – May 11	1
May 14 – May 18	5
May 21 – May 25	3
May 28 – June 1	2

How much did Carlos earn over the 5 weeks as a relief teacher?

8 Ashok is a casual office worker. He is paid $178 per day irrespective of the number of hours he works. Usually he works about 12 days per month.

a How much did Ashok earn for working from 8 a.m. to 1 p.m. on Monday.?

b During February, Ashok earned $1958. How many days did he work in February?

c The office offers Ashok a permanent 38-hour a week job at $16 per hour.
Do you think he should take the permanent job? Why or why not?

PS

9 Pia is trying to decide which one of 3 jobs to take.

PS

	Conditions	Pay
Job 1	38-hour week, 5 days per week, possibility of overtime	$19/hour
Job 2	75 hours per fortnight, work 9 days per fortnight	$1450 per fortnight
Job 3	Salary, based on a 35-hour week	$38 800 p.a.

a Ignoring any overtime, which job pays the most per year?

b If you were Pia, which job would you take? Why?

PROFILE

JONATHAN – SUPERMARKET MANAGER

When I was growing up there was never much money in my family.
If I wanted money, I had to work for it and I had lots of part-time jobs.
When I was in Year 9, I got an after-hours, part-time job stacking shelves in a supermarket. Because the manager was impressed with my work ethic and reliability, the company offered me a management trainee when I finished Year 12. Now I'm a supermarket manager with a range of responsibilities.

Imagefolk/Juice Images

Fair work

INVESTIGATION

AWARD WAGES

Search for 'A-Z Modern Awards' at the **Fair Work** website.

- Research the annual minimum wage for 3 jobs that interest you.
- Calculate the minimum weekly pay in each job.

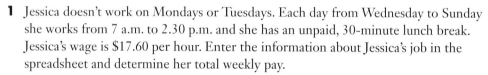

INVESTIGATION

WAGES BY SPREADSHEET

Wages

Ask your teacher to download the 'Wages' spreadsheet from the NelsonNet website.

1 Jessica doesn't work on Mondays or Tuesdays. Each day from Wednesday to Sunday she works from 7 a.m. to 2.30 p.m. and she has an unpaid, 30-minute lunch break. Jessica's wage is $17.60 per hour. Enter the information about Jessica's job in the spreadsheet and determine her total weekly pay.

2 a One of the formulas used in the spreadsheet is **=SUM(F11:F17)**. What is this formula calculating?

b What spreadsheet formula in cell F19 could be used to determine the total amount that Jessica is paid each week?

3 This table shows the hours worked during the first week in February and the corresponding pay rates for the employees in a small office.

Employee	Pay rate per hour	Number of hours worked	Pay
Imran	$12.51	20	
Sofia	$15.25	35	
Cathy	$20.70	35	
Mike	$16.30	40	
Anita	$16.30	32	
		Total wages bill	

a The hours each employee works per week and their hourly rate of pay could change. Construct a spreadsheet that will allow you to calculate each employee's wage and the total office wage bill when the number of hours worked and the rates of pay could change.

b During the second week in February each employee received a $4 per hour wage increase; Imran worked 32 hours and the other employees worked for the same number of hours as in the previous week. Use the spreadsheet you constructed to determine the total wages bill for the second week in February.

10.02 Working overtime

Overtime

Overtime is working beyond usual working hours, such as at night and on weekends, and is paid at a higher rate. Only people who work for a wage are paid for overtime (it doesn't apply to salary earners).

Overtime

Time-and-a-half is $1\frac{1}{2}$ times normal pay.

Double time is 2 times normal pay.

PROFILE

ALYSSA – AN AGED CARE WORKER

I helped my mum look after my grandfather who has dementia when I couldn't get a job after I left school. When we took Pa for a one-week holiday in a respite centre, the centre manager told me that I had the right attitude and I could consider working in the care service industry. I didn't need any special training because the employer provided on-the-job training and, with Australia's ageing population, there are lots of jobs to choose from. The best thing about my job is the variety. I deal with different clients and do different things every day.

EXAMPLE 3

Alyssa's normal rate of pay is $16.81 per hour. She is paid time-and-a-half on Saturdays and double time on Sundays.

a How much does Alyssa earn per hour on Saturdays?

b Calculate the amount that Alyssa will earn for working 4 hours on a Sunday.

Solution

a When she works on a Saturday, Alyssa is paid $1\frac{1}{2}$ times her normal rate.

Hourly pay on Saturdays $= 1.5 \times \$16.81$

$= \$25.22$

b When Alyssa works on Sunday, she earns $2 \times \$16.81$, or $33.62 per hour.

Pay for 4 hours on Sunday $= 4 \times \$33.62$

$= \$134.48$

Hasid earns $18 per hour.

a How much will he earn for working a 35-hour week?

b When Hasid works for more than 7 hours per day he is paid overtime. For the first 3 hours he works overtime, he is paid time-and-a-half. Any additional overtime hours are paid at double time. How much will Hasid earn for working 12 hours in one day?

Solution

a Multiply hourly rate by 35.

Pay for a 35-hour week $= 35 \times \$18$

$= \$630$

b When Hasid works for 12 hours in one day, his time is broken into 7 hours normal + 3 hours at time-and-a-half + 2 hours at double time.

Pay $= 7 \times \$18 + 3 \times 1.5 \times \$18 + 2 \times 2 \times \$18$

$= \$279$

Hasid's pay for a 12-hour day is $279.

Exercise 10.02 Working overtime

1 Complete the missing values in the table.

Normal pay per hour	Pay per hour at time-and-a-half	Pay per hour at double time
$17.20	a	b
$14.36	c	d
$24.60	e	f
$31.25	g	h

2 Jenny's normal pay is $16.40 per hour. How much will she earn when she works 5 hours at time-and-a-half?

3 Luis' normal pay is $15.30. How much will he earn when he works for 4 hours on a Sunday at double time?

4 How much will Sancia earn when she works 3 hours at time-and-a-half and 4 hours at double time? Her normal pay is $21.80 per hour.

5 Tuan is a plumber's assistant. He works a 35-hour week at $22.80 per hour. His overtime is paid at time-and-a-half for the first 5 hours overtime in a week and double time for any hours after that. This week Tuan worked 42 hours.

a How many hours did Tuan work at double time?

b Calculate Tuan's pay for the week.

6 Mercia has a holiday job supervising children in a resort. She earns $19 per hour Monday to Friday, time-and-a-half on Saturday and double time on Sunday.

a The table shows the times Mercia worked during one week in January. What are the missing values in the table?

Day	Hours worked	Pay rate per hour	Pay
Weekdays	21	i	iv
Saturday	4	ii	v
Sunday	6	iii	vi

b Calculate Mercia's pay for the week.

7 a Casey earned $108 when he worked for three hours at double time. What is Casey's normal pay per hour?

b How much will Casey earn when he works for 3 hours at time-and-a-half?

8 Elise earned $120 when she worked on Sunday for 4 hours at double time. How much does Elise earn for a normal 7-hour day?

9 Callum works for the council. He looks after the grass in parks and at sporting venues. Callum doesn't work any overtime on Monday to Friday. All the hours he works on Saturday are paid at time-and-a-half and Sunday work is at double time.
For Callum's time and pay sheet below, find the values of **a**, **b**, **c**, **d**, **e** and **f**.

Time and pay calculation sheet				
Callum O'Hare			Normal pay per hour **$16.90**	
Day	Start	Finish	Unpaid breaks	Pay
Monday	7 a.m.	3:30 p.m.	30 minutes	**a**
Tuesday	**b**	5:30 p.m.	1 hour	$152.10
Wednesday	–	–		
Thursday	8 a.m.	4 p.m.	**c**	$118.30
Friday	6:30 a.m.	**d**	1 hour	$135.20
Saturday	7 a.m.	11 a.m.	Nil	**e**
Sunday	8 a.m.	**f**	Nil	$101.40

10.03 Bonuses and allowances

Some jobs include **allowances** for doing unpleasant work, for working under difficult conditions, or to cover expenses such as uniform and travel.

Some jobs pay **bonuses** (extra pay) for doing good work, meeting targets or deadlines.

PROFILE

CAITLYN – CHEF IN THE AUSTRALIAN NAVY

I joined the navy because I didn't want a 9-to-5 job and I wanted to travel. I get good pay and conditions, as well as job security. I'm a fully-qualified chef and the navy provided all my training and arranged my TAFE qualifications. I've got good mates in the navy and I've been around the world. I was surprised at the variety of jobs in the navy; jobs I'd never considered: like being a waiter or a chaplain. The navy even has permanent jobs for musicians in the navy bands!

EXAMPLE 5

Caitlyn's basic salary in the navy is $43 434 and she receives an annual $12 128 service allowance as well as an annual $419 uniform maintenance allowance. When she's at sea she receives an additional $11 758 annually.

a   Calculate Caitlyn's weekly pay when she is working on land.

b   How much does Caitlyn earn per fortnight when she's at sea?

Solution

a   Caitlyn's annual salary on land = basic salary + service allowance + uniform allowance.

Salary = $43 434 + $12 128 + $419
= $55 981

Divide by 52 for weekly pay.

Weekly pay on land = $55 981 ÷ 52
≈ $1076.56

b   Caitlyn's annual salary at sea = basic salary + service allowance + uniform allowance + sea allowance.

Salary = $43 434 + $12 128 + $419 + $11 758
= $67 739

Divide by 26 for fortnightly pay.

Fortnightly pay at sea = $67 739 ÷ 26
≈ $2605.35

EXAMPLE 6

Sonia is paid $15.48 per hour as a security guard. Each week she receives an additional $61.05 for her guard dog and $6.75 for her torch. She receives $14.15 per shift travel allowance.

Sonia works a 4-hour shift, 6 nights per week. How much is she paid per week?

Solution

Sonia's total weekly pay	Wages $= 4 \times 6 \times \$15.48$
$=$ wages $+$ allowances $+$ dog $+$ torch	$= \$371.52$
	Travel allowance $= 6 \times \$14.15$
	$= \$84.90$
	Total weekly pay $= \$371.52 + \$84.90 + \$61.05$
	$+ \$6.75$
	$= \$524.22$

Exercise 10.03 Bonuses and allowances

1 Sophie's base salary as an air force trainee is $37 485 p.a. In addition, she receives the Australian Defence Force annual allowance of $12 128 and an annual $419 uniform allowance. She also receives $9531 p.a. when she is deployed overseas.

   a Calculate Sophie's weekly pay when she is working in Australia.

   b Determine Sophie's fortnightly pay when she is deployed overseas.

2 Zoran works for a pest control company. He is paid $14.93 per hour and he receives an extra $12.81 per day for handling poisons. Zoran works for 8 hours per day, 5 days per week. Calculate his weekly pay.

3 Ryan earns $721 per week as a mobile mechanic. In addition, he receives $29 per week for work-related phone calls and $0.60 per kilometre for work-related travel. Calculate Ryan's pay for a week in which he drove 420 km in his truck for work.

4 Kate is the manager of a fast food chain. She is paid $28 per hour for a 35-hour week plus a $8.30 per week laundry allowance. She receives a $30 bonus for every accident-free week at the shop and another $95 bonus if the shop makes $100 000 or more in sales. Last week the shop was accident-free and the sales were $110 000. How much was Kate paid last week?

5 Zack drives a furniture removal truck. He is paid $15.12 per hour Monday to Friday, time-and-a-quarter on Saturday and double time on Sunday. He receives a flat fee of $12.59 per day for handling heavy furniture. Calculate Zack's pay for a week when he delivered heavy furniture for 33 hours Monday to Friday, 6 hours on Saturday and 3 hours on Sunday.

PS **6** Raina has a job driving disabled children to school. She is paid $16.20 per hour plus $3.65 per day for assisting children. In addition, she receives 65 cents for every work-related kilometre she drives in her car. Calculate Raina's pay for a week when she worked 4 hours each day from Monday to Friday and she used her car for 360 work-related kilometres.

PS **7** Sam is a casual junior baker at the hot bread shop. A casual junior baker earns $12.32 per hour. From midnight Friday to midnight Saturday all bakers receive their normal pay plus 50%. From midnight on Saturday to midnight on Sunday casual bakers receive 98% more than their normal pay per hour.

a The table shows the times Sam worked last week. Complete the missing values in the table.

Shift	Starting time	Finishing time	Unpaid breaks	Number of hours worked	Pay per hour	Pay
1	Thursday 10 p.m.	Friday 6:30 a.m.	30 minutes	i	v	ix
2	Saturday midnight	8 a.m.	1 hour	ii	vi	x
3	Saturday 8 p.m.	Midnight	0	iii	vii	xi
4	Sunday 6:30 p.m.	Midnight	30 minutes	iv	viii	xii

b Calculate Sam's total pay.

My Future

Chapter problem

You've covered the skills required to solve the chapter problem. Can you solve it now?

10.04 Annual leave loading

Annual leave loading or **holiday loading** is an extra payment to employees given at the start of their holidays. It is usually calculated as $17\frac{1}{2}\%$ of 4 weeks' pay.

EXAMPLE 7

Briana earns $750 per week as a vet nurse. When she takes her 4 weeks annual holiday she receives an extra $17\frac{1}{2}\%$ of 4 weeks pay as holiday loading in addition to her normal 4 weeks pay.

a Calculate Briana's holiday loading.

b Determine the total value of Briana's holiday pay before she has to pay tax.

Solution

a Briana receives $17\frac{1}{2}\%$ of 4 weeks pay.

$$17\frac{1}{2}\% = 0.175 \longrightarrow$$

Pay for 4 weeks $= \$750 \times 4$

$$= \$3000$$

Briana's holiday loading $= 0.175 \times \$3000$

$$= \$525$$

b Holiday pay
= 4 weeks pay + holiday loading

Briana's holiday pay $= \$3000 + \$525$

$$= \$3525$$

Exercise 10.04 Annual leave loading

1 Calculate $17\frac{1}{2}$% of each amount.

   a $350   **b** $1264   **c** $3325   **d** $6895

2 The Edmondson Park Motel pays its employees a $17\frac{1}{2}$% leave loading on their 4 weeks annual leave.

   a Vicki, the chef, earns $695 per week. Calculate her annual leave loading.

   b James earns $565 per week as a barman at the hotel. How much will James be paid for his 4-week annual holiday?

3 Angus earns $743 per week and for his holidays he receives a loading of $17\frac{1}{2}$% of 4 weeks pay. Calculate the total value of his 4-week holiday pay.

4 As a result of an increase in the cost of living, all workers were granted a 4.2% increase in their pay.

   a Liam works as a data processing manager on a salary of $58 200 p.a. Calculate his new salary.

   b How much will Liam be paid for 4 weeks work after the wage rise?

   c Calculate Liam's new holiday loading.

5 Phillipa's annual salary is $72 320. She receives a loading of $17\frac{1}{2}$% of 4 weeks pay with her holiday pay. Calculate the total value of Phillipa's holiday pay.

6 Linda earns $890 per week. She receives 6 weeks holidays at the end of each year, but her holiday loading is only $17\frac{1}{2}$% of 4 weeks pay. Calculate Linda's holiday pay.

7 Jon's wage increased from $620 to $700 per week. By how much will his 4-week holiday pay, including $17\frac{1}{2}$% loading, increase?

8 PL Insurance had a very successful year. In addition to the normal $17\frac{1}{2}$% holiday loading, they decided to pay their employees a 'thank-you' bonus based on the number of years of service. They paid this bonus at the same time as the holiday loading.

Years of service	Bonus as a percentage of annual salary
1 – 5	0.4%
6 – 8	0.65%
Over 8	0.9%

Katrina is paid $2152 per fortnight and she has worked for the company for 7 years.

   a How much is Katrina's bonus?

   b Calculate the total amount that Katrina was paid, before tax, for her 4 weeks holiday, including the bonus.

9 Nate earns $640 per week. At the end of the year he receives 5 weeks holiday with a $17\frac{1}{2}\%$ loading on 4 weeks. How much more does Nate get paid for taking 5 weeks holidays than for working 5 weeks?

10 Ask your teacher to download the 'Holiday pay' spreadsheet from NelsonNet to answer this question.

Ask your teacher to download the 'Holiday pay' spreadsheet from NelsonNet to answer this question.

Holiday pay

a Yasmin earns $11.25 per hour for a 35-hour week. If the holiday leave loading increases from $17\frac{1}{2}\%$ to $22\frac{1}{2}\%$ of 4 weeks pay, by how much will Yasmin's 4-week holiday pay increase?

b What spreadsheet formula could be used to determine the value of:

   i normal pay for 1 week?

   ii normal pay for 4 weeks?

   iii annual leave loading for 4 weeks?

10.05 Commission, piecework and royalties

Earning money

Salespeople are often paid by **commission**, which is a percentage of the value of the items they've sold.

Piecework is a type of work where a worker is paid per item produced or processed.

A **royalty** is a payment to an author, singer or artist for each copy of their work sold. Usually a royalty is a percentage of the total sales amount.

Commission and piecework

EXAMPLE 8

Sarina receives a 5% royalty on the wholesale price of serviettes featuring her art. Packets of serviettes wholesale sell for $3.20 each. Calculate Sarina's royalty for the sale of 8000 packets of serviettes.

Solution

Find the value of the serviettes sold.

$$\text{Wholesale value} = 8000 \times \$3.20$$
$$= \$25\ 600$$

Calculate the royalty.

$$\text{Sarina's royalty} = 5\% \text{ of } \$25\ 600$$
$$= \$1280$$

EXAMPLE 9

Jordan is a used car salesman. He is paid a $170 monthly retainer plus 5% commission on his monthly sales over $50 000. Calculate his pay for a month when his sales totalled $80 000.

> A **retainer** is the amount of money a salesperson is paid that does not depend on his sales.

Solution

Find the value of the sales for which he is paid commission.

$$\text{Sales over } \$50\ 000 = \$80\ 000 - \$50\ 000$$
$$= \$30\ 000$$

Calculate the commission.

$$\text{Commission} = 5\% \text{ of } \$30\ 000$$
$$= \$1500$$

Total earnings = retainer + commission

$$\text{Total earnings} = \$170 + \$1500$$
$$= \$1670$$

EXAMPLE 10

Danielle earns commission for selling cosmetics, at the following rates:

Commission on monthly sales	
First $1000 of sales	5%
On the next $2000	4%
Remainder of sales	3.5%

These different rates are sometimes called a 'sliding scale'.

This month, Danielle's sales totalled $5200. Calculate her commission.

Solution

Danielle's sales figure can be broken into 3 sections:

$1000 + $2000 + remaining sales

Commission on the first $1000 = 5% × $1000

$$= \$50$$

Commission on the next $2000 = 4% × $2000

$$= \$80$$

Danielle's remaining sales = $5200 − $1000 − $3000

$$= \$2200$$

Commission on the remaining $2200 = 3.5% × $2200

$$= \$77$$

Calculate the total commission.

Danielle's total commission = $50 + $80 + $77

$$= \$207$$

Getty Images Plus/iStock/alexxx1981

EXAMPLE 11

Jan made some embroidered hand towels to sell at a charity street stall. To embroider the towels she bought:

- 10 plain hand towels at $8.25 each

- 8 metres of embroidery ribbon at 95 cents per metre

- 2 skeins of embroidery thread at $1.95

It took her $2\frac{1}{2}$ hours to make the towels and she values her time at $15 per hour. How much should Jan charge for her towels?

Solution

Find the total cost of the materials.

$$Cost = 10 \times \$8.25 + 8 \times \$0.95 + 2 \times \$1.95$$

$$= \$94$$

Remember to change the 95 cents into $0.95.

Find the cost of Jan's labour.

$$Labour = 2.5 \times \$15$$

$$= \$37.50$$

Add them together.

$$Total\ cost = \$94 + \$37.50$$

$$= \$131.50$$

Cost of one towel
= total cost ÷ number of towels

$$Cost\ per\ towel = \$131.50 \div 10$$

$$= \$13.15$$

A sensible price for Jan to charge for her towels should be a round number.

Jan should charge $13.50 or $14 per towel.

Getty Images./Moment Open/Lanzoni Matteo

Exercise 10.05 Commission, piecework and royalties

1 Calculate each percentage amount.

   a   9% of $25 000   **b**   5% of $800   **c**   2% of $300 000

   d $2\frac{1}{2}$% of $500 000   **e** $3\frac{3}{4}$% of $175 200   **f**   0.95% of $60 000

2 Marco earns 7% commission on all his sales. Find his commission on a sale of $1675.

3 Sarina receives a 3% royalty on the wholesale price of calendars featuring her art. How much royalty will she receive for 15 500 calendars with a wholesale price of $9.90?

4 Assam sells window shutters and is paid a retainer of $120 per week to cover his expenses, and a commission of 15% of all sales he makes. Assam's sales for the first week in April totalled $2896. Calculate his pay for that week.

5 In her job as a real estate agent, Lily is paid a retainer of $600 per month plus a commission of 2% of her sales over $800 000. How much did Lily earn for a month when her sales totalled $1 300 000?

6 Tanika sells cosmetics. She earns commission at the following rates.

Commission on monthly sales	
First $500 of sales	5%
On the next $1000	4%
Remainder of sales	3.5%

Calculate Tanika's commission for each monthly sales figure.

   a $360   **b** $1400   **c** $4200

7 Emily earns monthly commissions when she sells perfumes, according to these rates.

Monthly sales	Commission
$800	5% of sales
$801 to $1200	$40 plus 4.5% of sales over $800
$1201 and over	$58 plus 4% of sales over $1200

Calculate Emily's commission in a month when her total sales were valued at:

   a $360   **b** $998   **c** $5100.

8

Trevor's Tiling Service
Bathrooms, Kitchens

All Quality Work

* Tiles \$32 per m²
* Slate \$36 per m²
* 25% extra for stairs

Trevor told Monique that she required 6 m² of tiles for the kitchen walls, 48 m² of slate for the lounge room floor and 7 m² of slate for the stairs. How much will Trevor charge to lay the tiles and slate?

9 Ivana makes crystal pendants that she sells at the markets. To make 20 pendants it costs her:

- \$48 for the crystals

- \$55 for the chains

- \$10 for the clasps

Ivana values her time at \$18 per hour and it takes her 3 hours to make 20 pendants. What price should she charge for each pendant?

10 Holly enjoys cooking the scones she sells at the local Devonshire Tea shop. She buys her ingredients in bulk and it costs her \$12 to make 5 dozen scones in 2 hours. She values her labour at \$16 per hour. How much should Holly charge for making 10 dozen scones?

> 1 dozen = 12

11 Basam is selling his house for \$420 000. The real estate agent's commission is 2% on the first \$200 000 and 1.5% on the balance of the sale price. How much will Basam receive from the sale of his house?

12 Dagma writes books that sell for \$56 each. Each year she receives 10% royalty on the first 4000 copies and 12.5% royalty on the remaining sales of her books. Calculate the royalty Dagma receives in a year when:
  a 3650 copies of her books are sold
  b 7000 copies of her books are sold.

13 Milan stamps envelopes and posts letters for a marketing company. He is paid 24 cents per letter. Milan can process 70 letters per hour.
  a How much does Milan earn per hour?
  b How much will he earn for processing 260 letters?
  c How many letters does Milan need to process in order to earn over \$100?

14 Renuta is a furniture auctioneer. On every item she sells, she charges 15% commission on the first $2000 of the sale price and 12.5% of the amount remaining. How much will Renuta charge for selling an antique dining room suite that sold for $22 600?

15 The table shows the rates of royalty Sarina receives when her art is used on plates.

Number of plates sold	Royalty rate
First 2000	4% of the wholesale price
From 2001 to 10 000	$272 plus 3% of the wholesale price for the number of plates over 2000
10 001 or more	$1088 plus 2.5% of the whole price for the number of plates over 10 000

The wholesale price of the plates is $3.40. Calculate Sarina's royalty for the sale of each number of plates.

a 1500   **b** 8600   **c** 25 000

PROFILE

BEN – STUDENT

I have to look after my mum because there's just the two of us and she's got bipolar disorder. When she takes her medication she's OK, but I still need to do the shopping, cook our food and do the housework. It's really hard when I've got assignments and assessments, especially if Mum's unwell. I can't get a job because I have no spare time. Centrelink gives me a carer's pension and a carer's allowance. Mum receives sickness benefits and rent assistance.

INVESTIGATION

HELP FROM CENTRELINK

The Australian Government provides support for people in need. Visit the **Human Services** website and research the assistance available to people in Ben's situation.

Human services

10.06 Superannuation

How do people have enough income to pay for essential items and enjoy life when they retire? Things like cars and holidays can be very expensive and the government's Age Pension isn't enough to pay for luxuries.

Superannuation is a scheme that occurs while we're earning an income to save for our retirement years. Our employers invest 9.5% of our pay into our superannuation fund and we can make additional payments into the account if we want to.

EXAMPLE 12

Jake earns $24.80 per hour for a 35-hour week. His employer is required to pay 9.5% of his earnings into superannuation every 13 weeks. Calculate the amount of this superannuation payment.

Solution

Calculate Jake's pay for 13 weeks.

$$\text{Jake's weekly pay} = \$24.80 \times 35$$
$$= \$868$$
$$\text{Jake's pay for 13 weeks} = \$868 \times 13$$
$$= \$11\,284$$

Calculate 9.5% of 13 weeks' pay.

$$\text{Superannuation} = 9.5\% \times \$11\,284$$
$$= \$1071.98$$

Write the answer.

Jake's employer must pay $1071.98 into his superannuation account every 13 weeks.

On retirement, you will be able to access the money in your superannuation account, which can be paid to you as a lump sum or in regular instalments.

EXAMPLE 13

Katherine is retired and has 1.05 million dollars in her pension account created from her superannuation. Because she is 81 years old, the government requires that she withdraws 9% of the balance annually. Calculate the value of Katherine's monthly pension.

Solution

Calculate Katherine's annual pension.	Annual pension = 9% × \$1 050 000
	= \$94 500
Divide by 12 for the monthly pension.	Monthly pension = \$94 500 ÷ 12
	= \$7875
Write the answer.	Katherine's monthly pension is \$7875.

Exercise 10.06 Superannuation

Example
12

1 Zack earns \$29.15 per hour for a 38-hour week and his employer pays 9.5% of his pay into superannuation every 13 weeks. How much will Zack's employer pay into his superannuation account every 13 weeks?

2 Maryanne's annual salary is \$98 500.

   a Calculate the amount her employer must pay into superannuation every 13 weeks. Use a superannuation guarantee rate of 9.5%.

   b In 2025, the superannuation rate will increase to 12%. How much more will an employer have to pay into superannuation annually on a salary of \$98 500 p.a. compared to the 9.5% rate?

Shutterstock.com/VGstockstudio

3 Employers aren't required to pay superannuation for employees aged under 18 years unless they earn more than $450 per month and work more than 30 hours per week.

a Jane is 16 and works part-time at the local pizza store. This is part of her pay slip for February.

Dates	Hours worked	Pay
February 1 – 7	8	$121.60
February 8 – 14	7	$106.40
February 15 – 21	9	$136.80
February 22 – 28	12	$182.40
Superannuation		$0.00
Totals	**36**	**$547.20**

Jane is confused. She worked more than 30 hours and she earned more than $450. Why didn't she receive any superannuation?

b How much would the superannuation payment be if Jane was eligible for it?

c Research the minimum hourly rate of pay required to make a 16-year-old eligible for a superannuation payment.

4 Judith is 60 years old and has $784 000 in a superannuation account-based pension. She is required to withdraw at least 4% p.a. as a pension. Calculate Judith's pension if she decides to take the pension:

a annually   **b** monthly   **c** weekly

5 The minimum amount a person can take as a pension from a superannuation fund is determined by the person's age, as shown in the table.

a Matt is 71 years old and has $556 000 in a superannuation fund. He decided to only withdraw the minimum and to take a fortnightly pension. Calculate the amount of his fortnightly pension.

b Nanna has $967 500 in an account-based pension fund. She is about to have her 85th birthday. By how much will her annual pension change from age 84 to 85?

c Suggest a reason why the minimum pension percentages increase with age.

Age in years	Minimum percentage
Less than 65	4%
65 to 74	5%
75 to 79	6%
80 to 84	9%
85 to 89	10%
90 to 94	11%
95 or more	14%

INVESTIGATION

HOW MUCH WILL MY SUPERANNUATION BE WORTH?

Ask your teacher to download the 'How much will my superannuation be worth?' spreadsheet from NelsonNet.

Imagine that you've been working for 3 years and your employer has been contributing 9.5% of your wage into your superannuation account every 13 weeks. You've been saving hard and you've decided to go overseas to explore the world. Your superannuation account will continue to grow while you're away.

1 Enter a weekly pay rate of $700 in cell B7 and an interest rate of 5% (0.05) in cell B10.

Record how much your superannuation will be worth when you finish work in 3 years time and how much it will be worth in a further 10, 20, 30 and 40 years.

2 In which 10-year time period is the change in value the greatest?

3 Change your weekly pay. What effect does this have?

4 Change the interest rate. How does this affect the future value of your account?

5 Why is it important to start a superannuation account when you're young?

We will learn more about savings, investment and superannuation in Year 12.

KEYWORD ACTIVITY

Match each word in the first column to its correct meaning in the second column.

Word

Meaning

1	Allowance	**A**	Yearly
2	Annual leave loading	**B**	Pay based on the number of hours worked
3	Bonus	**C**	Pay based on the number of items made or processed
4	Double time	**D**	1.5 times the normal rate of pay
5	Income	**E**	A payment to authors, artists or others who create items
6	Overtime	**F**	Extra amount paid for holidays, usually 17.5% of 4 weeks' pay
7	Per annum (pa)	**G**	Extra pay for doing good work
8	Piecework	**H**	A fixed amount paid per year
9	Retirement	**I**	Twice the normal rate of pay
10	Royalty	**J**	Additional payment for work under difficult conditions or for doing unpleasant tasks
11	Salary	**K**	Money that is received or gained, usually regularly
12	Superannuation	**L**	Working more hours per day or week than normally
13	Time-and-a-half	**M**	A regular payment scheme for providing an income in retirement
14	Wage	**N**	The end of your formal working life

SOLUTION TO THE CHAPTER PROBLEM

Problem

Hugo earns $17.04 per hour for a 37-hour week in his job in a wholesale plant nursery. He is paid time-and-a-half for the first 5 hours of overtime per week and double time after that. If he is required to work any unscheduled overtime, he receives a $10.68 meal allowance per shift. In addition, he receives a $1.60 allowance per hour when he is required to work in wet areas.

Last week he worked 43 hours, which included one unscheduled overtime shift and 6 hours working in a wet area.

Hugo thinks his pay slip for the week ending 29 May is wrong. Is Hugo's gross pay correct?

Pay slip: Hugo Mendozia

Week ending 29 May

Normal pay	$630.48
Overtime	$153.36
Allowances	$20.28
Gross pay	$804.12

Solution

STAGE 1: WHAT IS THE PROBLEM? WHAT DO WE KNOW?

WHAT?

To find Hugo's pay to decide whether there's a mistake in his pay slip.

We know $17.04/h for 37 hours. Time-and-a-half for first 5 hours overtime, double time for any hours after that.

Meal allowance $10.68 for unscheduled overtime shift.

Wet allowance $1.60 per hour for 6 hours, Payslip shows $804.12

STAGE 2: SOLVE THE PROBLEM

SOLVE

Word clues: Gross → total

43 hours = 37 + 5 + 1

Hugo worked 37 hours normal time, 5 hours overtime at time-and-a-half and 1 hour at double time.

Normal pay	$37 \times \$17.04$	$630.48
Time-and-a-half	$5 \times 1.5 \times \$17.04$	$127.80
Double time	$1 \times 2 \times \$17.04$	$34.08
Total overtime	$127.80 + $34.08	$161.88
Wet area allowance	$6 \times \$1.60$	$9.60
Meal allowance		$10.68
Total allowances	$9.60 + $10.68	$20.28
Gross pay	$630.48 + $161.88 + $20.28	$812.64

The calculations for Hugo's normal pay and allowances are correct, but the overtime calculation is wrong.

Hugo has been underpaid by $161.88 − $153.36 = $8.52
(or $812.64 − $804.12 = $8.52).

STAGE 3: CHECK THE SOLUTION

Except for overtime, all the answers match the pay slip.

Check the overtime pay, $127.80 + $34.08 = $161.88.

Calculations are correct and complete.

CHECK

STAGE 4: PRESENT THE SOLUTION

Hugo has been underpaid by $8.52.

PRESENT

Earning money

1 Shelly earns $18.75 per hour for a 35-hour week.

   a How much does Shelly earn per week?

   b Calculate the amount she earns per fortnight.

   c What is Shelly's annual pay?

2 Marcus is an accountant. His annual salary is $96 000.

   a How much does Marcus earn per month?

   b Calculate Marcus' fortnightly pay.

3 Amal works in an aged-care facility. Her normal pay is $20.24 per hour.

   a How much does she earn per hour when she works at time-and-a-half?

   b When Amal works the night shift on weekends she is paid double time. Calculate her pay for working 7 hours on the late shift on a Saturday night.

4 George earned $168 for working 4 hours at double time. Calculate George's normal rate of pay.

5 Jamie is a plumber's assistant. His normal pay is $15.65 per hour, but when he has to work in wet or muddy conditions he receives a $2.10 allowance per hour. Jamie worked 7 hours today and for 3 of the hours he was digging in mud. Calculate Jamie's pay for today.

6 Angelo is a plumber. For emergency, late night call outs he charges a call out fee of $500 and labour at $125 per hour. Last night he received an emergency call at 2 a.m. Calculate the amount he will charge for going to the emergency and working for 3 hours to fix the problem.

7 Sabine is employed as an early childhood music teacher. Her normal weekly pay is $1325. Calculate her holiday pay for a 4-week holiday including a 17.5% leave loading.

8 Bart is an artist. His art features on fun park admission tickets. He receives 5% commission on the sale of each ticket. How much commission will Bart receive from the sale of 18 000 tickets priced at $45 each?

9 Luke lays tiles for a living. He charges $46/square metre of tiles he lays. How much will Luke charge for laying 17 square metres of tiles?

10 The balance on Brittany's superannuation-based pension account is $829 000 and she has chosen to take 5% of the balance annually as a fortnightly pension. Calculate the value of her fortnightly payment.

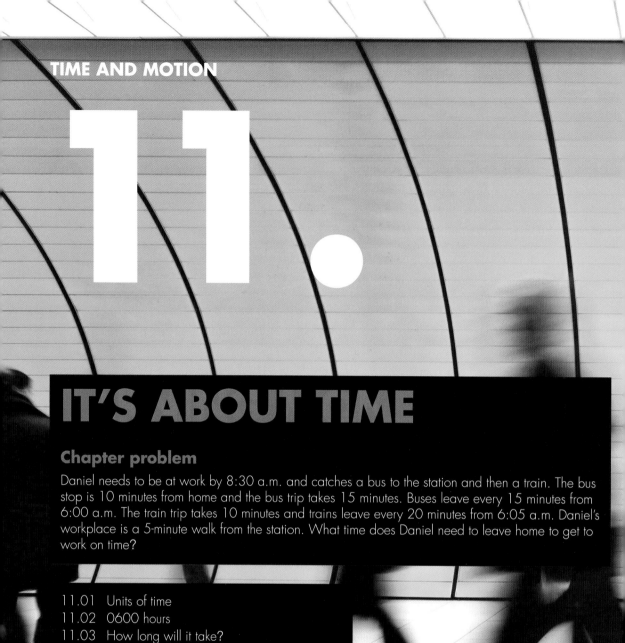

TIME AND MOTION

11.

IT'S ABOUT TIME

Chapter problem

Daniel needs to be at work by 8:30 a.m. and catches a bus to the station and then a train. The bus stop is 10 minutes from home and the bus trip takes 15 minutes. Buses leave every 15 minutes from 6:00 a.m. The train trip takes 10 minutes and trains leave every 20 minutes from 6:05 a.m. Daniel's workplace is a 5-minute walk from the station. What time does Daniel need to leave home to get to work on time?

WHAT WILL WE DO IN THIS CHAPTER?

- Convert between units of time: second, minute, hour, day, week, fortnight, month, year
- Write the time of day in 12- and 24-hour time and convert between the two formats
- Calculate time differences
- Solve problems involving Australian time zones
- Solve problems involving transport timetables and schedules, including planning trips

HOW ARE WE EVER GOING TO USE THIS?

- When looking up timetables and making plans for travel
- When checking TV programs or performance times
- Any time you need to calculate anything to do with time!

11.01 Units of time

Unit	Relationships
second (s)	
minute (min)	1 min = 60 s
hour (h)	1 h = 60 min = 3600 s
day	1 day = 24 h
week	1 week = 7 days
fortnight	1 fortnight = 2 weeks = 14 days
month	1 month = 30 or 31 days February has 28 days (or 29 days on leap years)
year	1 year = 12 months 1 year = 52 weeks 1 year = 365 days 1 leap year = 366 days

EXAMPLE 1

Convert:

a 23 minutes into seconds

b 91 days into weeks

Solution

a 1 min = 60 s, so multiply by 60.

$$23 \text{ min} = 23 \times 60 \text{ s}$$
$$= 1380 \text{ s}$$

b 1 week = 7 days, so divide by 7.

$$91 \text{ days} = 91 \div 7$$
$$= 13 \text{ weeks}$$

Scientific calculators have a degrees-minutes-seconds key ○′″ or DMS.

As we saw in Chapter 1, this key is useful for calculations involving hours, minutes and seconds.

Convert 275 minutes into hours and minutes.

Solution

1 min = 60 s so divide by 60.

275 min = 275 ÷ 60

= 4.583333… min

Press the ⓞ or **2ndF** **DMS** to change from decimal form to hours and minutes.

= 4 h 35 min

[4°35'0" on the calculator display]

Exercise 11.01 Units of time

1 Copy and complete each conversion.

Example 1

a 6 hours = _____ minutes

b $2\frac{1}{2}$ years = _____ weeks

c 480 seconds = _____ minutes

d 120 months = _____ years

e 96 hours = _____ days

f 3 fortnights = _____ days

g 5 years = _____ months

h 24 weeks = _____ fortnights

i 18 weeks = _____ days

j 78 months = _____ years

2 Saranya runs a marathon in $3\frac{1}{4}$ hours. What is this time in minutes?

3 Eddie is sentenced to spend 105 days in gaol. How many weeks is this?

4 Letitia is paid $3040 per fortnight. How much is this per week?

5 Yusuf takes out a loan to be paid back in monthly instalments over 15 years. How many monthly repayments is this?

6 Evan is paid $5585 per month.

a How much is Evan paid per year?

b How much is Evan paid per week?

7 Use the degrees-minutes-seconds key on your calculator to copy and complete each conversion.

Example 2

a 212 minutes = _____ hours _____ minutes

b 561 seconds = _____ minutes _____ seconds

c 330 seconds = _____ minutes _____ seconds

d 135 minutes = _____ hours _____ minutes

e 409 minutes = _____ hours _____ minutes

f 767 seconds = _____ minutes _____ seconds

8 In a training session, Zak takes 35 seconds to swim each 50 m lap.

   a How many seconds does Zak take to complete 40 laps?

   b Express your answer to part **a** in minutes and seconds.

   c Zak started swimming laps at 5 a.m. At what time will he finish the 40 laps?

9 Pooja is training for a marathon. She runs every day. On weekdays, she runs for $1\frac{1}{2}$ hours in the morning and $2\frac{1}{4}$ hours in the evening. On weekends, she runs 4 hours each day. For how long does Pooja run each week? Give your answer in hours and minutes.

10 Nelson Anglican College has a 20-minute break in the morning and a 45-minute lunch break. What is the total time for breaks in ← a fortnight? Give your answer in hours and minutes.

> How many school days in a fortnight?

11 In a charity marathon run, 7 people ran an equal amount of time in a 24-hour relay. Joanne was one of the runners. To work out how long she ran for, she did the following calculation:

$$24 \div 7$$

The calculator answer shows as 3.4285 …

   a Explain why the answer is NOT 3 hours 42 minutes.

   b Calculate the correct answer.

TECHNOLOGY

Calculating wages

On a spreadsheet, it is easier to enter times as decimals so they can be used in calculations.

1 Enter the following data into a spreadsheet.

	A	B	C	D	E
1	Calculating wages				
2					
3	Name	Time worked	Hours worked	Hourly rate	Wages
4	Grainger	37h 30 min	37.50	$27.40	$1,027.50
5	Sharwood	40h 12 min		$19.60	
6	Bush	39h 15 min		$25.90	
7	Nabaglo	29h 45 min		$32.20	
8	James	32h 30 min		$28.70	
9	Kerry	44h 36 min		$18.50	
10					
11			Total weekly wages =		
12					

2 Convert the times in column B to decimal hours and write the answers in column C. For Grainger, this should be 37.50 hours.

3 Enter the formula **= C5*D5** into cell E5.

4 Use **Fill Down** to complete column E.

5 Calculate the total weekly wages by entering **= SUM(E4:E9)** in cell E11.

11.02 0600 hours

There are two ways to show the time of day:

- 12-hour time: the conventional way using a.m. and p.m. and the hours 1 to 12
- 24-hour time: a more formal way that uses 4 digits, the hours 00 to 23 and no a.m./p.m.

There are two main types of clocks we use to tell the time:

- Analog clock with a clock face and the numbers 1 to 12 on it
- Digital clocks that can show either 12-hour or 24-hour time.

12-hour and 24-hour time

24-hour time on an analogue clock

Analog clock

Digital clock

This table shows the relationship between 12-hour and 24-hour time.

24-hour time	12-hour time	24-hour time	12-hour time
0000	12 a.m. (midnight)	1200	12 p.m. (midday)
0100	1 a.m.	1300	1 p.m.
0200	2 a.m.	1400	2 p.m.
0300	3 a.m.	1500	3 p.m.
0400	4 a.m.	1600	4 p.m.
0500	5 a.m.	1700	5 p.m.
0600	6 a.m.	1800	6 p.m.
0700	7 a.m.	1900	7 p.m.
0800	8 a.m.	2000	8 p.m.
0900	9 a.m.	2100	9 p.m.
1000	10 a.m.	2200	10 p.m.
1100	11 a.m.	2300	11 p.m.

EXAMPLE 3

Write 8:25 a.m. in 24-hour time.

Solution

8:25 a.m. is in the morning. Leave out the ':' and put a zero at the front

8:25 a.m. = 0825

EXAMPLE 4

Write 1615 as 12-hour time.

Solution

1615 is bigger than 1200 so it's in the afternoon (p.m.). For the hour, take 12 away from 16. 16 − 12 = 4.

1615 = 4:15 p.m.

Exercise 11.02 0600 hours

1 Write as 24-hour time:

   a   11:44 a.m.   **b**   6:35 p.m.   **c**   2:51 a.m.   **d**   9:54 p.m.

2 Write as 12-hour time:

   a   0845   **b**   1320   **c**   2331   **d**   1045

3 Write the time shown on each clock in 12-hour time:

   a    **b**    **c** 

4 Write the time shown on each clock in 12-hour time. All times are in the morning.

   a    **b**    **c** 

NELSON QMATHS 11. Essential Mathematics   ISBN 9780170412650

5 Write the time shown on each clock in 12-hour time. All times are in the afternoon or evening.

a

b

c

6 Show each time on an analog clock face.

a 8:15 a.m.

b 10:40 p.m.

c 3:35 p.m.

7 Copy and complete this table.

12-hour time	5 a.m.	a	2:10 p.m.	5:18 p.m.	b	c	d
24-hour time	e	0715	f	g	1730	2150	1120

8 Sue wants to program her DVR to record a television program when she's out. The program starts at 8:30 p.m. and lasts for an hour and a half. Her DVR works in 24-hour time.

a At what time will Sue need to set her DVR to start recording the program?

b What time should she set for her finish time? Allow an extra 5 minutes for the recording.

9 24-hour time is used in the army, navy and air force, so it is sometimes called 'military time'. Sean rings his wife from the army camp at 1730. What is this in 12-hour time?

10 Andrew was on police duty one Saturday night from 2200 to 0600 on Sunday morning. He receives a report of a fight at 2245 and again at 0220.

a How long was Andrew on duty?

b How long was it between the 2 incident reports?

c A patient is taken from the second fight to hospital and arrives 45 minutes after the incident was reported. What time does the patient arrive at hospital?

PS

11.03 How long will it take?

How long do I have before my favourite TV program starts? How long is my stopover in Hong Kong? How long have I worked here? We can answer all these questions by doing a time calculation.

EXAMPLE 5

What is the difference in time between 8:35 a.m. and 3:10 p.m.?

Solution

Use a timeline and count to the next full hour.

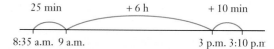

From 8:35 a.m. to 9:00 a.m.	25 minutes
From 9:00 a.m. to 3:00 p.m.	6 hours
From 3:00 p.m. to 3:10 p.m.	10 minutes
Add together.	Total time difference = 6 h + 25 min + 10 min
	= 6 h 35 min

EXAMPLE 6

What is the time 7 hours and 40 minutes after 11:52 p.m.?

Solution

Use a timeline and add the hours first:

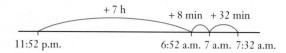

11:52 p.m. + 7 hours	6:52 a.m.
Count to the next full hour and add the number of minutes.	8 minutes to 7 a.m.
To make up 40 minutes, add another 32 minutes: 7 a.m. + 32 min.	7:32 a.m.

EXAMPLE 7

In March 2016, Kane's car was 5 years and 9 months old. When was Kane's car made?

Solution

Subtract 5 years from March 2016	$2016 - 5 = 2011$
	March 2011
Count back to the start of 2011 and subtract 3 months.	3 months to start of 2011
To make up 9 months, subtract another 6 months:	
Start of 2011 − 6 months	June 2010

Exercise 11.03 Time calculations

1 Calculate the time difference between each pair of times.

a 7:27 a.m. and 1:12 p.m. **b** 4:09 a.m. and 9:53 a.m.

c 3:42 p.m. and 6:02 p.m. **d** 11:15 p.m. and 3:08 a.m.

2 What is the time difference between each pair of 24-hour times?

a 0800 and 1100 **b** 0500 and 1500

c 0940 and 1455 **d** 1340 and 2150

3 Vamsee's flight lands in Hong Kong at 0950. His next flight leaves Hong Kong at 1920. How long does Vamsee have in Hong Kong airport?

4 At Thomson Secondary College, the school day begins at 8:45 a.m. and ends at 3:10 p.m. How long is the school day?

5 What time will it be:

a 5 hours after 3:00 p.m.? **b** 28 minutes after 7:15 p.m.?

c 3 hours 32 minutes after 9:45 a.m.? **d** 9 hours 10 minutes after 5:14 p.m.?

6 What time was it:

a 4 hours before 6:15 p.m.? **b** 45 minutes before 3:20 a.m.?

c 2 hours 10 minutes before 1:35 p.m.? **d** 3 hours 35 minutes before 11:25 a.m.?

7 Ewan has to take his tablets $1\frac{1}{2}$ hours after he finishes lunch. He finished his lunch at 1.45 p.m. What time should Ewan take his medicine?

8 Ramona wants to go for a jog for 50 minutes, but be back in time to watch *Zombie Vampire Athletes* on TV starting at 8:35 p.m. What is the latest time that she can leave for her jog?

9 Jamie bought a computer in December 2015. How old is it:

    a   in August 2019?   **b**   in January 2020?   **c**   today?

10 A box of chocolates bought in November 2018 has a 'Use by' date of March 2020. How long, in months, will the chocolates last?

11 In February 2018 Stephanie was $3\frac{1}{2}$ years old. When was Stephanie born?

12 Ask your teacher when they started teaching at your school. Work out how long they have been at your school in years, months and days.

13 How old are you today in years, months and days?

11.04 Times across Australia

Australia has 3 time zones:

- **Australian Eastern Standard Time (AEST)**, covering Queensland, New South Wales, the ACT, Victoria and Tasmania

- **Australian Central Standard Time (ACST)**, covering South Australia and the Northern Territory and $\frac{1}{2}$ hour behind AEST

- **Australian Western Standard Time (AWST)**, covering Western Australia and 2 hours behind AEST

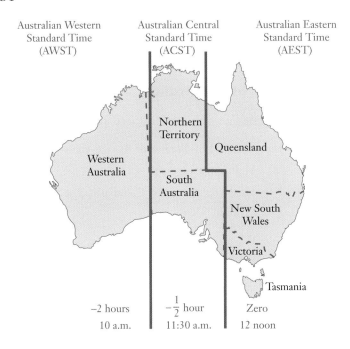

| −2 hours | −$\frac{1}{2}$ hour | Zero |
| 10 a.m. | 11:30 a.m. | 12 noon |

EXAMPLE 8

Show the Australian time zones on a timeline with Western Australia as 0 hours.

Solution

If Western Australia is 0, then Eastern Australia is + 2 hours and Central Australia is $+1\frac{1}{2}$ hours.

0		+2
AWST	ACST	AEST
0	$+1\frac{1}{2}$	+2

When moving east, add on the time difference.

When moving west, subtract the time difference.

ISBN 9780170412650

EXAMPLE 9

a It is 9 a.m. in Perth. What time is it in Adelaide?

b It is 2 p.m. in Brisbane. What time is it in Adelaide?

Solution

a Moving east from Perth to Adelaide:
add the time difference.

$9 \text{ a.m.} + 1\frac{1}{2} \text{ hours} = 10:30 \text{ a.m.}$

b Moving west from Brisbane to Adelaide:
subtract the time difference.

$2 \text{ p.m.} - \frac{1}{2} \text{ hour} = 1:30 \text{ p.m.}$

Exercise 11.04 Times across Australia

1 Show the time zones on a timeline with the Northern Territory as 0 hours.

2 State whether each location is ahead of, behind or has the same time as Adelaide.

 a   Brisbane   **b**   Melbourne   **c**   Darwin   **d**   Canberra

 e   Geraldton (WA)   **f**   Mt Isa (Qld)   **g**   Ballarat (Vic)   **h**   Ceduna (SA)

 i   Hobart   **j**   Broome (WA)

3 It is 10:30 a.m. in Adelaide. What time is it in:

 a   Cairns (Qld)?   **b**   Freemantle (WA)?   **c**   Alice Springs (NT)?

 d   Launceston (Tas)?   **e**   Bendigo (Vic)?   **f**   Broome (WA)?

4 What is the time in each city when it is 11 p.m. in Brisbane?

 a   Melbourne   **b**   Adelaide   **c**   Perth

 d   Darwin   **e**   Hobart   **f**   Canberra

5 The AFL Grand Final started in Melbourne at 2:15 p.m. and was shown live in Adelaide. What time did the game start on Adelaide TV?

6 a   Mick's flight from Brisbane to Perth will take $5\frac{1}{2}$ hours. He leaves Brisbane at 2 p.m. What time does he land in Perth? Give your answer as Perth local time.

 b   When Mick flies home, he leaves Perth at 11:30 a.m. What time does he land in Brisbane? Give your answer as Brisbane local time.

7 When Joanna was holidaying at Margaret River in WA, she phoned her parents in Cairns. She rang at 6:30 p.m. Western Standard Time. What time was it in Cairns?

8 Simone, in Toowoomba, uses the Internet to talk with her cousins in Fremantle, WA. At what time should Simone log on to catch her cousins at 8:00 p.m. Fremantle time?

9 a Use the Internet to find out when daylight saving begins and ends in NSW and Victoria.

b Why do some states have daylight saving?

c How does daylight saving affect the different time zones?

d When it is 12:30 p.m. in Western Australia (not on daylight saving),what time is it in NSW on Eastern Daylight Saving Time?

10 Broken Hill in NSW operates on Central Standard Time rather than Eastern Standard Time like the rest of NSW. Suggest a reason why Broken Hill might have chosen to do this. (Hint: look at a map.)

11 Karl is watching the cricket at the Melbourne Cricket Ground live on TV at 2.30 p.m., but the clock at the cricket ground says 5.30 p.m. (Eastern Daylight Saving Time). Where is Karl watching from?

12 Isabella and Lara like to talk on Facetime. On New Year's Eve, Lara called at 10 p.m. local time and it was 7.30 p.m. in Isabella's location. Where could the girls be located?

11.05 Timetables

Thousands of Australians rely on rail, bus and ferry **timetables** every day.

Time travel

EXAMPLE 10

A section of a train timetable is shown here.

Armadale/Thornlie to Perth

TV times

Source: Transperth

Armadale	7:16 AM		7:46 AM		8:16 AM		8:31 AM	
Sherwood	7:18 AM		7:48 AM		8:18 AM		8:33 AM	
Challis S	7:20 AM		7:50 AM		8:20 AM		8:35 AM	
Kelmscott	7:22 AM		7:52 AM		8:22 AM		8:37 AM	
Seaforth	7:25 AM		7:55 AM		8:25 AM		8:40 AM	
Gosnells	7:27 AM		7:57 AM		8:27 AM		8:42 AM	
Maddington	7:30 AM		8:00 AM		8:30 AM		8:45 AM	
Kenwick	7:32 AM		8:02 AM		8:32 AM		8:47 AM	
Beckenham	7:34 AM		8:04 AM		8:34 AM		8:49 AM	
Thornlie		7:35 AM		8:05 AM		8:35 AM		8:50 AM
Cannington	7:36 AM	7:39 AM	8:06 AM	8:09 AM	8:36 AM	8:39 AM	8:51 AM	8:54 AM
Queens Park		7:40 AM		8:10 AM		8:40 AM		8:55 AM
Welshpool		7:42 AM		8:12 AM		8:42 AM		8:57 AM
Oats Street	7:39 AM	7:44 AM	8:09 AM	8:14 AM	8:39 AM	8:44 AM	8:54 AM	8:59 AM

Public transport trips

Carlisle		7:45 AM		8:15 AM		8:45 AM		9:00 AM
Victoria Park		7:48 AM		8:18 AM		8:48 AM		9:03 AM
Burswood		7:50 AM		8:20 AM		8:50 AM		9:05 AM
Belmont Park				8:22 AM		8:52 AM		9:07 AM
Claisebrook	7:47 AM	7:54 AM	8:17 AM	8:24 AM	8:47 AM	8:54 AM	9:02 AM	9:09 AM
Mciver	7:49 AM	7:56 AM	8:19 AM	8:26 AM	8:49 AM	8:56 AM	9:04 AM	9:11 AM
Perth	7:50 AM	7:58 AM	8:20 AM	8:28 AM	8:50 AM	8:58 AM	9:05 AM	9:13 AM

Perth to Armadale/Thornlie

Perth	4:37 PM	4:45 PM	4:52 PM	5:00 PM	5:07 PM	5:15 PM	5:22 PM	5:30 PM
Mciver	4:38 PM	4:46 PM	4:53 PM	5:01 PM	5:08 PM	5:16 PM	5:23 PM	5:31 PM
Claisebrook	4:40 PM	4:48 PM	4:55 PM	5:03 PM	5:10 PM	5:18 PM	5:25 PM	5:33 PM
Belmont Park	4:42 PM		4:57 PM		5:12 PM		5:27 PM	
Burswood	4:44 PM		4:59 PM		5:14 PM		5:29 PM	
Victoria Park	4:46 PM		5:01 PM		5:16 PM		5:31 PM	
Carlisle	4:48 PM		5:03 PM		5:18 PM		5:33 PM	
Oats Street	4:49 PM	4:54 PM	5:04 PM	5:09 PM	5:19 PM	5:24 PM	5:34 PM	5:39 PM
Welshpool	4:51 PM		5:06 PM		5:21 PM		5:36 PM	
Queens Park	4:53 PM		5:08 PM		5:23 PM		5:38 PM	
Cannington	4:55 PM	4:58 PM	5:10 PM	5:13 PM	5:25 PM	5:28 PM	5:40 PM	5:43 PM
Beckenham		4:59 PM		5:14 PM		5:29 PM		5:44 PM
Kenwick		5:01 PM		5:16 PM		5:31 PM		5:46 PM
Maddington		5:03 PM		5:18 PM		5:33 PM		5:48 PM
Gosnells		5:07 PM		5:22 PM		5:37 PM		5:52 PM
Seaforth		5:09 PM		5:24 PM		5:39 PM		5:54 PM
Kelmscott		5:12 PM		5:27 PM		5:42 PM		5:57 PM
Challis		5:14 PM		5:29 PM		5:44 PM		5:59 PM
Sherwood		5:16 PM		5:31 PM		5:46 PM		6:01 PM
Armadale		5:19 PM		5:34 PM		5:49 PM		6:04 PM
Thornlie	5:00 PM		5:15 PM		5:30 PM		5:45 PM	

Louise catches the 7:52 a.m. train from Kelmscott to get to work at Burswood. What time does she arrive at Burswood station?

Solution

The 7:52 a.m. train from Kelmscott does not stop at Burswood. She will need to change trains at Oats Street. She can then catch a train to Burswood.

The train to Burswood arrives at the station at 8:20 a.m.

Exercise 11.05 Timetables

Use the timetable on the previous pages to answer each question.

Example 10

1 What time does the 7:16 a.m. train from Armadale arrive at Perth?

2 What time does the 5:03 p.m. train from Carlisle arrive at Thornlie?

3 Ravi lives at Queens Park and works in Perth. He starts work at 8:45 a.m. What is the latest train he can catch at Queens Park to get to work on time?

4 Ravi finishes work at 4:30 p.m. It takes him 10 minutes to walk to the station.

   a What time can he catch the train home?

   b What time does he get to Queens Park?

   c Three days a week Ravi catches up with friends after work for half an hour.
On these days, what time does he catch the train home?

5 Frances catches the 5 p.m. train from Perth. What time will she arrive at Gosnells?

6 Katarina is going to catch the train at Maddington to meet her friend Vimala at Belmont Park. They have agreed to meet for shopping and coffee at 9:15 a.m. What time does Katarina need to catch the train?

7 It takes 1 hour and 10 minutes to travel by bus from Thornlie to Perth.

   a How much longer does it take by bus than by train?

   b Suggest reasons why the bus might take longer.

8 Use the Internet to find a local or Brisbane train timetable. Save it or print a copy.

   a Write 5 questions, similar to those in questions **1** to **7**, for the timetable you have found.

   b Swap questions with a friend and answer their questions.

This is the bus timetable for Adelaide to Alice Springs and return. Use it to answer questions **9** to **13**.

ALICE SPRINGS - ADELAIDE

TOWN	CODE		GX850 DAILY EST	PICK-UP AND SET-DOWN POINT
ALICE SPRINGS	ASP	Dep	10.30A	Greyhound Terminal, Shop 2/76 Todd St
Erldunda	ERL		12.35P	Desert Oaks Resort, Cnr Stuart & Lassester Hwy
Kulgera	KUL	Arr	1.25P	Kulgera Roadhouse Hotel, Stuart Hwy
		Dep	2.00P	
SOUTH AUSTRALIA				
Indulkna Turnoff	IDK		3:20P	Roadside mailbox at turn off
Marla	MBR		4:00P	Traveller's Rest, Stuart Hwy
Cadney Park	CDY		4:55P	Mobil Roadhouse
Coober Pedy	CPD	Arr	6:30P	Terminal, 52-56 Hutchison St
		Dep	7:25P	
Bilgunnia Turnoff	BGN		8:55P	Turn Off
Glendambo	GBO	Arr	10:10P	Mobil Roadhouse
		Dep	10:30P	
Pimba	PIM	Arr	11:45P	Pimba Roadhouse
		Dep	12:01A	
Port Augusta	PUG		1:50A	Post Office, 50 Commerical Road
Port Augusta Meal Break	PUB	Arr	2:00A	Gull Service Station, Lot 8 Highway
		Dep	2:45A	
Port Pirie	PIR	Arr	3:40A	BP Service Station
Port Wakefield	PWF	Dep	5:00A	BP Service Station, 26 Snowtown Rd.
Bolivar	BLV		5:45A	BP Service Station
Cavan	CAV		6:00A	Bus Stop 26, Port Wakefield Road
ADELAIDE	ADL	Arr	6:25A	Greyhound Terminal, 85 Franklin St

ADELAIDE - ALICE SPRINGS

TOWN	CODE		GX580 DAILY EST	PICK-UP AND SET-DOWN POINT
ADELAIDE	ADL	Dep	6.00P	Greyhound Terminal, 85 Franklin St
Cavan	CAV		6.20P	Bus Stop 26, Port Wakefield Road
Bolivar	BLV		6.30P	Caltex Service Station
Port Wakefield	PWF		7.25P	BP Service Station, 26 Snowtown Rd.
Port Pirie	PIR		8.40P	BP Service Station

ISBN 9780170412650

Port Augusta Meal Break	PUB	Arr	9.45P	Gull Service Station, Lot 8 Highway
		Dep	10.30P	
Port Augusta	PUG		10.50P	Post Office, 50 Commerical Rd
Pimba	PIM	Arr	12.45A	Shell Roadhouse
		Dep	1.00A	
Glendambo	GBO	Arr	2.15A	Shell Roadhouse
		Dep	2.30A	
Bulgunnia Turnoff	BGN		3.35A	Turn Off
Coober Pedy	CPD	Arr	5.15A	Terminal, 52-56 Hutchison St
		Dep	5.50A	
Cadney Park	CDY	Arr	7.30A	Mobil Roadhouse
Marla	MBR		8.20A	Traveller's Rest, Stuart Hwy
			9.05A	
Indulkna Turn Off	IDK		9.35A	Roadside mailbox at turn off
NORTHERN TERRITORY				
Kulgera	KUL		11.00A	Kulgera Roadhouse Hotel, Stuart Hwy
Erldunda	ERL	Arr	11.45A	Desert Oaks Resort, Cnr Stuart & Lassester Hwy
		Dep	12.20P	
ALICE SPRINGS	ASP	Arr	2.30P	Greyhound Terminal, Shop 2/76 Todd St

Greyhound Australia, with permission.

9 How long does the trip from Adelaide to Alice Springs take?

10 How many times during the trip does the bus stop for more than 20 minutes? Where are these stops?

11 Harry joins the bus at Port Augusta to travel to Coober Pedy.
  a How long is his bus trip?
  b When Harry returns from Coober Pedy he will travel all the way to Adelaide. How long will it take?

12 Ariana and Axel are friends. Ariana lives in Adelaide and Axel lives in Alice Springs. They are planning to meet 'in the middle' to spend the weekend together.
  a Where is the closest stop to 'the middle'?
  b Taking into account the time of day and when each person would arrive, where would be the best place to meet along this route? Justify your answer.

13 The train from Adelaide to Alice Springs leaves Adelaide at 12:20 p.m. Sunday and arrives in Alice Springs at 1:45 p.m. Monday.
  a How long does it take to travel by train from Adelaide to Alice Springs?
  b How does this time compare to the bus trip?
  c Would you prefer to travel by bus or train from Adelaide to Alice Springs? Justify your answer.

This timetable is for CityCat ferries operating on the Brisbane River. Use it to answer questions **14** to **16**.

Departs Terminal:	a.m.	a.m.	a.m.	a.m.	a.m.	a.m.	a.m.	a.m.	p.m.
Northshore Hamilton	.....	.....	.....	.....	.....	.....	.....	.....	.....
Apollo Wharf	10.25	10.37	10.50	11.02	11.15	11.27	11.40	11.52	12.05
Bretts Wharf	10.28	10.40	10.53	11.05	11.18	11.30	11.43	11.55	12.08
Bulimba	10.34	10.46	10.59	11.11	11.24	11.36	11.49	12.01	12.14
Teneriffe	10.37	10.49	11.02	11.14	11.27	11.39	11.52	12.04	12.17
Hawthorne	10.41	10.53	11.06	11.18	11.31	11.43	11.56	12.08	12.21
New Farm Park	10.46	10.58	11.11	11.23	11.36	11.48	12.01	12.13	12.26
Mowbray Park	10.50	11.02	11.15	11.27	11.40	11.52	12.05	12.17	12.30
Sydney Street	10.53	11.05	11.18	11.30	11.43	11.55	12.08	12.20	12.33
Riverside	11.01	11.13	11.26	11.38	11.51	12.03	12.16	12.28	12.41
QUT Gardens Point	11.09	11.21	11.34	11.46	11.59	12.11	12.24	12.36	12.49
South Bank 2	11.12	11.24	11.37	11.49	12.02	12.14	12.27	12.39	12.52
North Quay	11.16	11.28	11.41	11.53	12.06	12.18	12.31	12.43	12.56
Regatta	11.24	11.36	11.49	12.01	12.14	12.26	12.39	12.51	1.04
Guyatt Park	11.28	11.40	11.53	12.05	12.18	12.30	12.43	12.55	1.08
West End	11.31	11.43	11.56	12.08	12.21	12.33	12.46	12.58	1.11
University of Queensland	11.35	11.47	12.00	12.12	12.25	12.37	12.50	1.02	1.15

https://jp.translink.com.au/plan-your-journey/timetables/ferry/t/citycat

14 How long does the ferry take to travel from Bretts Wharf to South Bank 2?

15 a Phillip needs to be at work in a restaurant at Riverside by 11:30 a.m. What is the latest time he can catch the ferry at Teneriffe?

b After work he travels out to the University of Queensland to meet up with friends. How long does the ferry trip take?

c At the end of the evening he returns home. How long will this trip take if he chooses to go by ferry? Assume the reverse journey will take the same length of time as the forward journey.

d To travel from the University of Queensland to Teneriffe by bus takes approximately 55 minutes. Why might people choose to travel by bus rather than ferry?

16 Why is one part of the timetable grey and one part unshaded?

17 The City Explorer Bus stops at places of interest in the city and operates in Brisbane, Sydney and Melbourne. This is the timetable for the explorer bus in Brisbane.

GPO terminal	9:00	9:45	10:30	11:15	12:00	12:45	1:30	2:15	3:00	3:45	4:30	5:15
City Hall	9:05	9:50	10:35	11:20	12:05	12:50	1:35	2:20	3:05	3:50	4:35	5:20
Treasury Casino Hotel	9:10	9:55	10:40	11:25	12:10	12:55	1:40	2:25	3:10	3:55	4:40	5:25
Riverside Centre	9:15	10:00	10:45	11:30	12:15	1:00	1:45	2:30	3:15	4:00	4:45	5:30
Old Windmill	9:20	10:05	10:50	11:35	12:20	1:05	1:50	2:35	3:20	4:05	4:50	5:35
Transit Centre	9:28	10:13	10:58	11:43	12:28	1:13	1:58	2:43	3:28	4:13	4:58	5:43
Suncorp Stadium	9:32	10:17	11:02	11:47	12:32	1:17	2:02	2:47	3:32	4:17	5:02	5:47
Regatta Hotel	9:45	10:30	11:15	12:00	12:45	1:30	2:15	3:00	3:45	4:30	5:15	6:00
Park Rd	9:48	10:33	11:18	12:03	12:48	1:33	2:18	3:03	3:48	4:33	5:18	6:03
Cultural Centre	9:54	10:39	11:24	12:09	12:54	1:39	2:24	3:09	3:54	4:39	5:24	6:09
Southbank	10:00	10:45	11:30	12:15	1:00	1:45	2:30	3:15	4:00	4:45	5:30	6:15
Maritime Museum	10:04	10:49	11:34	12:19	1:04	1:49	2:34	3:19	4:04	4:49	5:34	6:19
City lookout	10:10	10:55	11:40	12:25	1:10	1:55	2:40	3:25	4:10	4:55	5:40	6:25
Chinatown	10:20	11:05	11:50	12:35	1:20	2:05	2:50	3:35	4:20	5:05	5:50	6:35
ANZAC Square	10:24	11:09	11:54	12:39	1:24	2:09	2:54	3:39	4:24	5:09	5:54	6:39
GPO terminal	10:30	11:15	12:00	12:45	1:30	2:15	3:00	3:45	4:30	5:15	6:00	6:4

Source: http://theaustralianexplorer.com.au/

a How many buses are needed to meet the Explorer Bus timetable? Explain how you arrived at your answer.

b Vo, Binh and Vicki arrived at City Hall at 11:42 a.m. They caught the Explorer Bus to Southbank. What is the earliest time they could expect to arrive at Southbank? Explain your answer.

c Manuel and Sofia are dropped off by car at the Riverside Centre at 10:25 am. They arrange to meet their hosts at the City Lookout at 4:30 p.m. They want to spend at least an hour at Suncorp Stadium, ride on the Ferris Wheel at Southbank and do some souvenir shopping at the Maritime Museum. Plan a list of times for them to catch the Explorer Bus to do these things and meet their hosts on time.

d The company is considering introducing an early morning bus starting at the GPO terminal at 8.15 a.m. Write out the timetable for this bus.

PS

ISBN 9780170412650

PLANNING A TRIP

You are visiting one of the state capitals in Australia outside of Queensland. You are going to plan a trip from the airport to the CBD and on to an island off the coast.

1 Choose which capital you are going to visit.

2 You need to travel from the airport to the CBD for your accommodation. Choose a hotel in the capital city to stay at. Find out the best way to travel from the airport to the hotel, how long it will take and what it will cost.

3 You are going to visit one of the islands near the city you have chosen (Rottnest near Perth, Kangaroo near Adelaide or Phillip near Melbourne, for example). Find out how to get there from your hotel (train, bus and/or ferry), how long it will take and what it will cost.

4 Choose another tourist attraction in the capital city and find out how to get there from your hotel.

5 Find out if there any special public transport deals available in the capital city you have chosen that could save you money when you are visiting the city.

11.06 Nature's timetables

Tide charts tell us when high and low tide are along our coastline. There are sunrise and sunset charts, and there are also charts for the phases of the moon.

EXAMPLE 11

This chart shows information about tides (time and height) and sunrise/sunset times for Cairns over two weeks in October.

Date	High tides				Low tides				Sunrise/sunset	
October	a.m.	m	p.m.	m	a.m.	m	p.m.	m	a.m.	p.m.
Sun 13	4:59	2.02	5:36	2.76	10:50	0.85	11:54	0.77	5:49	6:16
Mon 14	5:53	2.24	6:22	2.85	11:43	0.72			5:49	6:16
Tue 15	6:40	2.45	7:04	2.88	12:30	0.60	12:30	0.63	5:48	6:16
Wed 16	7:24	2.60	7:40	2.84	1:05	0.48	1:12	0.60	5:47	6:17
Thu 17	8:03	2.70	8:11	2.73	1:39	0.41	1:51	0.64	5:46	6:17
Fri 18	8:39	2.74	8:38	2.58	2:13	0.39	2:30	0.74	5:46	6:17
Sat 19	9:13	2.71	9:04	2.40	2:46	0.43	3:08	0.89	5:45	6:17
Sun 20	9:48	2.63	9:31	2.19	3:18	0.53	3:46	1.08	5:45	6:18
Mon 21	10:25	2.50	9:57	1.96	3:50	0.68	4:26	1.29	5:44	6:18

Tue 22	11:05	2.33	10:12	1.75	4:21	0.87	5:10	1.49	5:43	6:18
Wed 23	11:55	2.17			4:49	1.06			5:43	6:19
Thu 24	2:48	2.09			5:19	1.24			5:42	6:19
Fri 25	4:15	1.44	3:57	2.17	1:28	1.35	6:08	1.41	5:42	6:19
Sat 26	4:42	2.26			12:12	1.28	11:56	1.18	5:41	6:20

The following time adjustments may be made for other places near Cairns.

Flinders Island: 22 min after Cairns

Cape Flattery: 10 min before Cairns

Morris Island: 14 min after Cairns

Cape Grenville: 51 min after Cairns

Cruiser Pass: 7 min before Cairns

Michaelmas Cay: 11 min before Cairns

Euston Reef: 27 min before Cairns

Swallows Landing: 2 min after Cairns

Find the time of the second high tide on 17 October in:

a Cairns   **b** Morris Island   **c** Euston Reef

Solution

a Find Oct 17 in the main table, and then the second high tide.

8:11 p.m.

b High tide at Morris Island is 14 min after Cairns.

8:11 p.m. + 14 min = 8:25 p.m.

c High tide at Euston Reef is 27 min before Cairns.

8:11 p.m. − 27 min = 7:44 p.m.

Exercise 11.06 Nature's timetables

1 Use the tide table on page 284 the time of the morning low tide on 21 October in:

    a   Cairns   **b**   Cruiser Pass   **c**   Swallows Landing

2 Find the time of the first high tide on 25 October in:

    a   Cairns   **b**   Flinders Island   **c**   Cape Flattery

3 a   On 22 October, calculate how many hours and minutes are between the 2 low tides.

    b   What is the height difference in metres between these low tides?

4 a   On 15 October what are the times for sunrise and sunset?

    b   How many hours of daylight are there on this day?

5 Most days have 4 tides. Which days have fewer than 4 tides?

6 a   Which day has the earliest sunrise?

    b   What is happening to the time of sunrise over these 14 days?

7 Jacob likes to fish on the 'rising tide', when the tide is changing from low to high.

    a   Approximately when could he go fishing at Cairns on 19 October?

    b   The next day he decides to fish at Cape Grenville. Approximately when could he go?

8 At Michaelmas Cay, boats cannot be launched within one hour of low tide because the water is too shallow. Kylie wants to launch her boat on the afternoon of 16 October. After what time can she launch the boat?

9 a   How many hours of daylight are there on 26 October?

    b   What is happening to the hours of sunlight over these 14 days?

10 There also tables for the phases of the moon. Choose a city in Queensland and find a table of moon phases for this city.

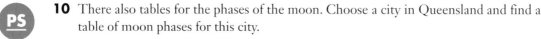

INVESTIGATION

MY NEXT TRIP

You are planning to experience a uniquely Australian attraction for you next holiday:

- see the fairy penguins at Phillip Island at dusk   OR

- see the sun rise over Uluru.

Choose which attraction you will see and plan your trip. Create a PowerPoint or a display book that shows all times, methods of transport and costs.

Share your plans with the rest of the class.

TRAVELLING TO FRASER ISLAND

1 Investigate online the options for travelling to Fraser Island from where you live. These may be by car, bus, train and/or plane. Google Maps may help.

- Write down all the options and how long each one takes.

- Give a reason why you might choose each of the different options. Consider factors such as time taken, cost, activities you might do on the way, reasons for travelling.

2 Choose two places that you could visit, one in Queensland, one in another part of Australia, and repeat Question 1.

KEYWORD ACTIVITY

Describe each term below in a sentence and match it with one of the photos below:

12-hour time   24-hour time   timeline   time zone   timetable

A

B

C

D

E

SOLUTION ^{TO} _{THE} CHAPTER PROBLEM

Problem

Daniel needs to be at work by 8:30 a.m. and catches a bus to the station and then a train. The bus stop is 10 minutes from home and the bus trip takes 15 minutes. Buses leave every 15 minutes from 6:00 a.m. The train trip takes 10 minutes and trains leave every 20 minutes from 6:05 a.m. Daniel's workplace is a 5-minute walk from the station. What time does Daniel need to leave home to get to work on time?

Solution

STAGE 1: WHAT IS THE PROBLEM? WHAT DO WE KNOW?

To work out what time Daniel needs to leave home to arrive at work at 8:30 a.m. There is a lot of information given about bus, train and walking times.

WHAT?

STAGE 2: SOLVE THE PROBLEM

We need to work backwards from 8:30 a.m.

SOLVE

Daniel needs to be at the station at least 5 minutes before 8:30 a.m, that is, at 8:25 a.m. (5 minutes to walk)

Daniel must catch the train by 8:15 a.m. (10-minute train trip)

Trains leave at 6:05 a.m. and then every 20 minutes, so we need to calculate train departure times: 6:05, 6:25, 6:45, 7:05, 7:25, 7:45, 8:05 and 8:25.

Daniel must catch the train at 8:05 a.m.

Daniel must catch the bus by 7:50 a.m. (15-minute bus trip)

Buses leave at 6:00 a.m. and then every 15 minutes, so the bus departure times are 6:00, 6:15, 6:30, 6:45, 7:00, 7:15, 7:30, 7:45 and 8:00.

Daniel must catch the bus at 7:45 a.m.

Daniel must leave home by 7:35 a.m. (10-minute walk to bus stop)

CHECK

STAGE 3: CHECK THE SOLUTION

Daniel leaves home at 7:35, catches the bus at 7:45, arrives at the train station at 8:00, catches the train at 8:05, arrives at 8:15, walks to work and arrives at 8:20, which is before 8:30, so this solution works.

Depending on the reliability of the bus and train, he may choose to leave earlier to be sure to get to work on time.

PRESENT

STAGE 4: PRESENT THE SOLUTION

Daniel needs to leave home at or before 7:35 a.m.

11. CHAPTER REVIEW

It's about time

1 Copy and complete each conversion.

    a 720 seconds = _____ minutes   **b** 7 weeks = _____ days

    c $4\frac{1}{2}$ years = _____ months   **d** 192 hours = _____ days

2 a Arima runs 8 km in 31 minutes and 17 seconds. What time is this in seconds?

    b The *Franken Furniture* Super Sale runs for 96 hours. How many days is this?

3 a Express in 24-hour time:

    i 4:17 a.m.   **ii** 3:25 p.m.   **iii** 10 a.m.

    b Write in 12-hour time:

    i 0615   **ii** 1640   **iii** 2300

4 a A bus leaves Brisbane at 0945 and arrives at Rockhampton at 2120. How long is the bus trip?

    b Janine works as a paramedic. Her shift starts at 2300 and ends at 0600.
She has 4 callouts during her shift – at 2325, 0048, 0215 and 0450.

      i How long is her shift?

      ii How long was it between the last 2 callouts?

      iii The callout at 0048 required transport to hospital which took 25 minutes.
What time did Janine arrive at the hospital?

5 a Calculate the time difference between 11:15 a.m. and 3:40 p.m.

    b What time was it 3 hours and 5 minutes before 2:30 p.m.?

6 a Lalaja has a stopover at New Delhi airport from 1015 to 1505. How long will she be at the airport?

    b Keith wants to record a film screening on TV for 2 hours 40 minutes.
He sets the recorder to start at 8 p.m. What time should he set it to finish?

    c In January 2017, Maya had been paying off her car monthly for 22 months.
In what month and year did she start paying off the car?

7 Use the map of Australian time zones on page 275 to answer these questions.

    a Name one city beyond Queensland that is behind the time where you live.

    b If it is 11.30 a.m. in Perth, WA, what time is it Brisbane?

    c If it is 3.15 p.m. in Innisfail, Qld, what time is it in Adelaide, SA?

8 Use the train timetable from Example 10 on page 277 to answer these questions.

   a Anthony catches the 8.25 a.m. train from Seaforth to Perth. What time does he arrive in the city?

   b The Kane family are travelling from Gosnells to Victoria Park to catch up with friends. They have planned to meet at 9 a.m. What train should they catch from Gosnells?

   c The Kane family want to be home by 5.30 p.m. What train should they catch from Victoria Park?

9 Use the bus timetable from Exercise 11.05 on page 280 to answer these questions.

   a Carmel is travelling from Alice Springs to Coober Pedy. How long will her trip take?

   b Mick joins the bus at Port Augusta to travel to Marla. What time will he arrive in Marla?

   c On the trip from Alice Springs to Adelaide the bus has long stops in Kulgera, Coober Pedy and Port Augusta. Give a possible reason for these longer breaks.

10 Use the ferry timetable from Exercise 11.05 on page 282 to answer these questions.

   a How long does the ferry take to travel from New Farm Park to the University of Queensland?

   b Nate lives near Apollo Wharf and starts work near Riverside at 12 noon. What is the latest time he can catch the ferry and not be late?

11 Use the tide chart from Example 11 on page 284 to answer these questions.

   a At what time is the morning low tide at Michaelmas Cay on 19 October?

   b How many hours of daylight are there on 24 October?

   c On October 16, what is the difference in hours and minutes between the 2 high tides?

   d What is the height difference in metres between these high tides?

12.

CENSUS AND SURVEYS

Chapter problem

Andrew is a market researcher. The United Club has asked him to evaluate the facilities and services it provides for members. Should he survey all 14 100 members or take a sample? If he chooses to use a sample, how should he select it to ensure that he has a good representative range of views?

WHAT WILL WE DO IN THIS CHAPTER?

- Follow the process of statistical inquiry
- Learn the difference between a census and a sample
- Examine the various types of samples and how to choose them

HOW ARE WE EVER GOING TO USE THIS?

- When we read and interpret statistical information in the media
- When we undertake a survey
- When we evaluate the results of surveys

12.01 Census vs sample

Statistics involves the collecting of information, which is then analysed and used to make decisions. To collect information, we usually survey a representative group. This process is called taking a **sample**.

To collect information about a whole population, *all* people or items must be surveyed. This is called taking a **census**. People who are members of small groups can be missed in a sample. We always use a census when we want to make sure that the views of small groups are included.

The Australian Bureau of Statistics conducts a national census every 5 years. This census is held in years ending in a 1 or a 6, such as 2016.

Sample	Census
• Surveys a selected group of people or items	• Surveys all people or items in the population
• Gives approximate information about the population	• Gives exact information about the population
• Simple and inexpensive	• Complex and expensive
• Can be done quickly	• It takes a lot of time to collect and process the information

EXAMPLE 1

The United Club asked Andrew to evaluate the facilities and services it provides to its 14 100 members. Should he use a census or a sample to gather information?

Solution

A census would be expensive and it would take a long time to process the information.

A sample should be used in this situation.

EXAMPLE 2

From what population should Andrew take a sample to investigate the reputation of the United Club in the local community?

Solution

Population means the people Andrew should be asking.

The population would be the residents of the community where the United Club is located.

Exercise 12.01 Census vs sample

1 Write a paragraph in your own words describing the advantages and disadvantages of using a census to collect information.

2 For each investigation, should a census or a sample be used? Give a reason for your answer.

a The most popular car colour in Australia

b The number of retired people living on the Sunshine Coast

c The number of Australians who watch the NRL Grand Final

d The use of soap versus body wash

e Testing coffee for taste

f The population of Winton

g The number of people using the emergency department at Redcliffe Hospital on Saturday night

h Length of time a certain type of car battery lasts

i The political party supported by people in Queensland

j The favourite TV show of your 10 best friends

k Steve Smith's batting average

l Internet usage by Year 11 students at your school

m Hours of paid work per week completed by senior students in your school

n Nelson Pay TV wants to know how satisfied its customers are with its service

3 What is the target population for each investigation?

a Girls are better than boys at Maths

b Voting intentions for the next State election

c The best song of the last decade

d Student attitudes to school uniform at your school

e Favourite make of car in Queensland

f The amount of money people are prepared to spend on going to the gym each week

g The venue Year 12 should use for their formal

h Donations to charities from wealthy individuals

i The factors influencing the choice of supermarket

j The batting performance of the Australian cricket team

k NQM Bank wants to know if its customers find Internet banking easy to use

l The time taken to deal with complaints to a phone company

m The local council wants to know what recreational equipment should be added to the town's parks

Australian Bureau
of Statistics

Australian
statistics

Census
questions

THE AUSTRALIAN CENSUS

The Australian census is conducted by the **Australian Bureau of Statistics**.
Visit its website and search for the answers to these questions.

1 When was the first national census conducted?

2 When was the last census?

3 List 5 questions that were asked in the last census.

4 Find 3 questions that have been asked in the past but were not asked in the most recent census.

5 Were there any questions in the most recent census that have not been asked before? If so, what were they?

6 Are all questions in the Census compulsory?

7 Give 3 examples of how census information is used.

8 What was different about the Census in 2016? What challenges did this present?

9 Who is NOT required to complete the Census?

12.02 The statistical investigation process

Information is collected by a variety of groups to help people make informed decisions. This process is called **statistical investigation** and it involves the following steps:

- **Posing questions** – deciding what information is needed and in what form

- **Collecting data** – choosing between using a **census** or a **sample** of the population, deciding how we will collect information and then collecting the information

- **Organising data** –organising the collected information for better studying, such as using a frequency table, with either grouped or ungrouped data

- **Summarising and displaying data** – presenting the information in a way that makes it easy to follow and understand, such as with graphs or tables

- **Analysing data** – calculate summary statistics such as mean, median and mode, then looking for patterns and relationships in the data

- **Writing a report** – presenting the results of the investigation in a clear way, with the conclusions supported by the statistics

Exercise 12.02 The statistical investigation process

1 In your own words, summarise the steps in the process of statistical investigation.

2 Classify each activity below as one of these steps:
- posing questions (PQ)
- collecting data (CD)
- organising data (OD)
- displaying data (DD)
- analysing data and drawing conclusions (AD)
- writing a report (WR)

a Liong asks fellow students in Year 11 how they travel to school

b Jane draws a frequency histogram of her data

c Anna finds the mean house price for her suburb

d Managing Director Theo decides he needs to know customers' favourite car colour

e Kieran makes recommendations in his final report

f Emilia puts her information on favourite holiday destinations into a frequency table

g Lee concludes that coffee is her friends' favourite daytime drink

h Simon visits the residents of Lawson Street to collect their Census forms

i Madeline decides to research the drinking habits of 18-year-olds

j Will displays data about pocket money received by Year 11 students in a sector graph

k Phoebe writes a report on her data about popular sports

l Kim goes through a pile of surveys and records the responses to a specific question

3 Choose a topic to research and describe how you would implement each of the 6 steps of a statistical investigation.

PS

4 Some organisations collect information on a large scale. Visit the websites of each organisation listed and write at least 5 things they collect information about.

a the Australian Bureau of Statistics (ABS)

b the United Nations (UN)

c the World Health Organization (WHO)

Australian Bureau of Statistics

United Nations

5 The privacy of collected information can be important. Visit the ABS website and search for the **Census** page. Scroll down and click on **Privacy, Confidentiality & Security**. Describe in your own words how the ABS ensures the privacy of the information we provide in the Census.

World Health Organisation

12.03 Types of samples

If we decide to use a sample for a survey, there are 4 types of samples we can use.

In a **random sample**, every member of the population has an equal chance of being included in the sample. For example, when a computer selects a customer of a phone company at random to be surveyed about the company's performance, every customer of that company has an equal chance of being selected.

In a **stratified sample**, different categories in a population are represented according to their proportion size of the population and then members of each category are selected randomly.

For example, a school's population may be made up of 72% junior students and 28% senior students. A stratified sample of school students must also be made up of 72% junior students and 28% senior students.

In a **systematic sample**, selections are made on a regular basis. For example, testing every 500th battery produced in a factory to check that the machines producing the batteries are working properly.

In a **self-selected sample**, whoever wishes to participate answers the questions asked. For example, when a current affairs TV program asks people to vote online about who should be Australia's next prime minister, anyone can participate.

EXAMPLE 3

The United Club has 14 100 members. This table gives a breakdown of members by age group.

Andrew is considering 4 different of ways of choosing a sample for his market research. Which type of sample is each one?

Age group	Number of members
Less than 30 years	2100
30 to 39 years	2700
40 to 49 years	5100
50 years and over	4200

a   Every 100th member on the alphabetical membership list

b   Names selected randomly by the computer

c   Putting up a sign at the entrance of the club asking members to volunteer

d   21 members who are less than 30 years old, 27 members who are 30 to 39 years old, 51 members who are 40 to 49 years old and 42 members aged 50 years or more.

Solution

a   Selections are made on a regular basis, in this case every 100th member.   Systematic sample

b   Every member of the population has an equal chance of being included and the computer chooses randomly.   Random sample

| c | Any member who volunteers can participate. | Self-selected sample |
| d | A predetermined number of members are chosen based on the number of members in each age group. | Stratified sample |

Exercise 12.03 Types of samples

1 Which type of sampling is described in each case?

   a Selecting every 105th name from the phone book

   b The names of all Year 11 students are placed in a hat and 2 are drawn out to represent the school at a council function

   c A television program asks viewers to respond to a Yes/No question

   d Selecting an appropriate number of students from each year at your local high school

   e Subscribers are sent an email asking for their opinion of the play they have just seen

   f Names of employees at a bank are drawn out of a barrel

   g The audience at a concert finds prize tickets under every 40th seat in each row

   h A company sends out an email to customers asking for their opinions

   i Employees are sorted from tallest to shortest and every 5th employee completes a questionnaire

   j 20 females and 28 males were surveyed out of a group of athletes with 100 females and 140 males

   k An airport customs officer searching every 10th airline passenger

   l A medical researcher advertises for people to participate in a health survey

   m 5 cards are selected from a pack of cards without looking

   n An import/export business employs 125 women and 250 men. 17 women and 34 men are surveyed about their work hours

   o A computer selects every 1000th name from the electoral (voters') roll

   p A bank surveys 5% of its customers in each age group

2 Children from 400 families attend the local primary school. The P&C has raised money for new play equipment for the playground and wants to interview parents about their ideas. The committee would like to survey 40 families. Suggest how they might select the 40 families using a:

   a random sample

   b stratified sample

   c systematic sample

   d self-selected sample

3 For the survey above, which method of sampling is the best to use? Explain your answer.

4 What are the possible disadvantages of using:

   a a self-selected sample?

   b a random sample?

   c a systematic sample?

   d a stratified sample?

Chapter problem

You've covered the skills required to solve the chapter problem. Can you solve it now?

INVESTIGATION

SAMPLE SIZES IN POLLS

In this investigation, we will look at the sizes of samples used by polling organisations.

Roy Morgan

1 Visit the **Roy Morgan** research website.

2 Click on the **Findings** tab and select one of the polls in Australia that interests you.

3 Write down what the poll is about and the size of the sample.

4 Use the Internet to find the population of Australia.

5 What percentage is the sample of Australia's population?

Remember: $\dfrac{\text{sample}}{\text{population}} \times 100$

Galaxy Research

6 Visit the **Galaxy Research** website.

7 Select a recent national or state poll and write down the sample size they used.

8 Use the Internet to find the population of Australia or the state for the poll you have chosen.

9 What percentage is the sample of the population?

10 Write 2 or 3 sentences describing what you have found.

11 How well do you think the samples reflect the opinions of the population?

12.04 Choosing a sample

When we have decided what type of sample to use, we need to know how to choose the sample.

EXAMPLE 4

Selina decided to use a stratified sample to survey members at her local gym. There are 970 gym members, of which 590 are female and 380 are male. She plans to survey 10% of the members.

a How many members will Selina survey?

b How many female members should she survey?

c How many male members should she survey?

Solution

a Calculate 10% of the membership.

10% of 970 = 97

Selina should survey 97 members.

b Female members make up 590 out of 970 members.

Write this as a fraction and find this fraction of the 97 members to be surveyed.

$\dfrac{590}{970} \times 97 = 59$

Selina should survey 59 female members

c Male members make up 380 out of 970 members.

OR, from part b.

$\dfrac{380}{970} \times 97 = 38$

97 members – 59 females = 38 males

Selina should survey 38 male members.

Exercise 12.04 Choosing a sample

1 Maria wishes to survey the members of the fan club she belongs to. Suggest any method she could use to find a random sample of members to survey.

2 Mr Carrozza, the school principal, wants to survey a sample of students' parents at Nelson Valley College. Suggest how he could find a systematic sample of parents.

3 In a self-selected sample, people volunteer to complete a survey. Suppose you need to collect data on usage of the school canteen. Suggest 3 places in your school where you might make questionnaires available for those who wanted to fill them in.

4 Amanda is going to use a stratified sample to survey the parents of her local netball club. There are 540 children playing netball, made up of 380 primary school students and 160 high school students. Amanda is going to survey 15% of parents.

   a How many parents should complete the survey?

   b How many parents of primary students should complete the survey?

   c How many parents of high school students should complete the survey?

5 Kieran is going to survey Year 8 students about their opinions of the school. There are 96 boys and 69 girls in Year 8. He aims to survey 20% of students.

   a How many students should complete the survey?

   b How many boys should complete the survey?

6 Jamiela is going to ask a sample of her Facebook friends about their favourite sport. She has 376 single friends and 416 friends in a relationship. She is going to survey 12.5% of her friends.

   a How many friends will she survey?

   b How many single friends will be in the survey?

   c How many friends in a relationship will be in the survey?

7 Global Communications employs 750 people made up of 479 males and 271 females. The company intends to survey 75 employees about their working conditions.

   a How many males should be surveyed?

   b How many females should be surveyed?

8 Darren is researching political views in Queensland. He stands on Roma Street in central Brisbane and interviews every 10th person who walks past.

   a Is this a representative sample of the population of Queensland? Why or why not?

   b If not, describe how Darren could achieve a representative sample.

WORD MATCH

Census and surveys
find-a-word

Match each word with its definition.

1	census	**A**	A representative group in a survey
2	population	**B**	Every member of the population has an equal chance of being selected
3	privacy	**C**	Different categories in a population are represented according to the size of each category
4	questionnaire	**D**	Your right not to share personal information
5	random	**E**	The process of collecting information to enable people to make informed decisions
6	sample	**F**	One of the most common ways to collect information
7	self-selected	**G**	Sample where selections are made on a regular basis, such as every 100th item
8	stratified	**H**	The group of people from whom a sample is chosen
9	statistical investigation	**I**	Whoever wishes to participate answers the questions asked
10	systematic	**J**	All people or items are surveyed

SOLUTION TO THE
CHAPTER PROBLEM

Problem

Andrew is a market researcher. He has been asked to evaluate the facilities and services being provided to the community by the United Club. Should he survey all 14 100 members or take a sample? If he chooses to use a sample, how should he select it to ensure that he gets a good representative range of views?

Solution

STAGE 1: WHAT IS THE PROBLEM? WHAT DO WE KNOW?

We need to decide who Andrew will survey and what type of sample he could use.

WHAT?

STAGE 2: SOLVE THE PROBLEM

It would be too expensive to survey all 14 100 members and it would take too long to analyse the results. A sample is the better option.

SOLVE

To ensure a representative sample, Andrew should take a stratified sample of the different age groups. If he is using a sample size of 10%, then he should survey 10% of each age group.

STAGE 3: CHECK THE SOLUTION

This solution makes sense and we have answered both parts of the question.

CHECK

STAGE 4: PRESENT THE SOLUTION

Andrew should take a stratified sample of the members of the club, surveying 10% of members in each age group.

PRESENT

12. CHAPTER REVIEW

Census and surveys

1 Should you use a census or a sample to find the most popular TV program in Queensland? Give a reason for your answer.

2 For each investigation listed below:

    i what is the target population?

    ii should a census or sample be used?

  a The number of children in each family in your street

  b Make of car in car fleets of Queensland companies

  c Voting intentions at the next federal election

  d Whether boys complete more homework than girls in Queensland schools

3 List the steps in a statistical investigation in your own words.

4 Cengage Engineering has 430 employees. The CEO wishes to survey the employees about working conditions at the company. He would like to survey 86 employees. Suggest how the firm contracted to do the research might select the 86 employees using a:

  a random sample

  b stratified sample

  c systematic sample

  d self-selected sample

5 Lindsay is going to use a stratified sample to survey the parents of all students in her dance classes. There are 375 children in the dance classes: 295 girls and 80 boys. Lindsay is going to survey 20% of parents.

  a How many parents should complete the survey?

  b How many parents of girls should complete the survey?

  c How many parents of boys should complete the survey?

Practice set 3 ●●●○

Section A Multiple-choice questions

For each question, select the correct answer **A**, **B**, **C** or **D**.

1 What is the time difference between 10:42 a.m. and 2:13 p.m.?

   A 3 h 31 min   **B** 4 h 55 min

   C 8 h 29 min   **D** 12 h 55 min

2 Joshua creates 3 graphs showing the information he found in his survey. What step is this in the statistical investigation process?

   A Collecting data   **B** Organising data

   C Representing data   **D** Analysing data

3 Sunil earns $3672 per month. What is his weekly salary?

   A $1836   **B** $1694.77

   C $918   **D** $847.38

4 For which situation would you use a census rather than a sample to investigate?

   A Testing coffee for the best taste

   B Finding the number of migrants from Japan in Australia

   C Opinions on road safety in Queensland

   D People's views on whether Australia should become a republic

5 Ursula lives in Brisbane and rings her friend in Perth at 1:15 p.m. What time is it in Perth?

   A 11:15 a.m.   **B** 12.45 p.m.

   C 1:45 p.m.   **D** 3:15 p.m.

6 Regina sells cars and is paid a weekly retainer of $320 plus 3.5% of the value of the cars she sells. Calculate her income for a week when she sells cars worth $87 000.

   A $331.20   **B** $1120

   C $3045   **D** $3365

7 A radio station asks listeners to complete an online survey about their favourite music. What sort of sample is this?

A random   **B** self-selected

C stratified   **D** systematic

8 Jeremy works 12 hours at a normal pay rate of $22.80 per hour and 3 hours overtime at time-and-a-half. Which calculation could be used to find his weekly pay?

A $22.80 × 15   **B** $22.80 × 15 × 1.5

C $22.80 × 12 + $22.80 × 3 × 1.5   **D** $22.80 × 12 + $22.80 × 15 × 1.5

Section B   Short-answer questions

1 Copy and complete each statement.

  a   4 weeks = _____ days

  b   84 months = _____ years

  c   108 hours = _____ days

  d   22 weeks = _____ fortnights

2 The Australian Bureau of Statistics conducts a census every 5 years. Give an example of how the data it collects is summarised and displayed.

3 a   Michaela is paid an annual salary of $109 000. How much is this each fortnight?

  b   Tyson is paid $33.50 per hour. How much is he paid for a 35-hour week?

4 Vincent wants to record a movie on TV. The program says the movie starts at 21:30 and runs for $2\frac{1}{2}$ hours.

  a   Convert 21:30 to 12-hour time.

  b   Allow an extra 15 minutes for the recording, in case the movie starts and finishes late. What time should he set for his finish time? Answer in 12-hour time.

5 Melinda wants to survey student views on the school uniform. Describe how she could choose each type of sample.

  a   random sample

  b   stratified sample

  c   self-selected sample

  d   systematic sample

Exercise
10.02

6 Simon is paid $276 for working 6 hours on a Sunday at double time. What is his usual hourly rate of pay?

Exercise
11.05

7 Use the train timetable from Example 10 on page 277 to answer these questions.

   a Chantelle catches the 7:57 a.m. train from Gosnells to go to work near Oats Street. At what time does she arrive at Oats Street?

   b Scott lives in Sherwood. He goes to Cannington to work. He is allowed to arrive any time between 7 a.m. and 9 a.m. At what times could he catch the train?

   c Denise catches the 5:07 p.m. train from Perth. At what time can she arrive at Seaforth?

Exercise
10.04

8 Anilha earns $970 per week. Each year, she is paid 17.5% annual leave loading on 4 weeks pay. Calculate:

   a her pay for 4 weeks

   b the annual leave loading she will be paid

   c the total amount she earns for her 4-week holiday

Exercise
12.01

9 a Give an example of an investigation where you would use a census to collect the information.

   b Who would be the target population for your investigation?

Exercise
11.04

10 a A football match starts in Brisbane at 4 p.m. What time is this in:

    i Adelaide?

    ii Perth?

   b An AFL match starts in Darwin at 11.30 a.m. What time is this in:

    i Cairns?

    ii Fremantle, WA?

11 Melinda decides to use a stratified sample of 10% of students to survey their views on school uniform. She has the following data from the school office.

Exercise
12.04

Total students	840
Male	390
Female	450

Year 7	165
Year 8	128
Year 9	171

Year 10	155
Year 11	127
Year 12	94

a How many students should complete the survey?

b How many female students should be part of the survey?

c How many Year 8 students will be surveyed?

12 Judy and Sue are paid royalties on the sale of their books. Each author is paid 7.5% of the selling price of the book. Calculate the royalties each author will receive on the sale of 2400 books at $49.95 each.

Exercise
10.05

13

7.20	3 hours 0 min	12.345
DISTANCE	TIME	STEPS

HEALTHY FIGURES

Chapter problem

For morning tea, Renee had a chocolate muffin (825 kJ) and some coffee with milk and sugar (295 kJ). Afterwards, she goes for a walk, using 23 kJ of energy per minute. How long will Renee have to walk to use all the energy she consumed at morning tea?

Movement

LOW

01   02   03   04   05   06

WHAT WILL WE DO IN THIS CHAPTER?

- Convert between kilojoules (kJ) and calories (Cal) units of energy
- Apply units of energy and rates to solve a variety of problems involving food, exercise and electricity

HOW ARE WE EVER GOING TO USE THIS?

- Exercising adequately to maintain a healthy weight
- Consuming an appropriate level of food for our nutritional needs

13   14   16   17

Shutterstock.com/Africa Studio

13.01 Burning energy

Kilojoules (kJ) are metric units of energy. If we use more energy than we eat, we lose body weight. However, if we eat more energy than we use, this energy is stored as fat.

This table shows the amount of energy our body needs each day, depending on our age group, sex and level of activity.

Age	Lifestyle	Men kJ/day	Women kJ/day
18–35	inactive	10 500	8000
	active	12 500	9000
	very active	14 800	10 500
36–70	inactive	10 000	8000
	active	11 800	8800
	very active	14 300	10 400
Pregnant women			10 100
Breast-feeding women			11 800

EXAMPLE 1

Kylie is 18 years old, works in an office and has an inactive lifestyle. Every day on her way to work she has a cappuccino at the coffee shop. The cappuccino contains 940 kJ.

a How many kilojoules of energy does Kylie need each day?

b What percentage of her daily energy needs does Kylie have in her morning cappuccino?

Solution

a Kylie is 18, a female, and has an inactive lifestyle. Her energy needs are shown in the top row and right column in the table.

8000
Kylie needs 8000 kJ per day.

b $\dfrac{\text{kJ in the cappuccino}}{\text{kJ per day}} \times 100\%$

$\dfrac{940}{8000} \times 100\% = 11.75\%$

Kylie's cappuccino contains 11.75% of her daily energy needs.

Calories (Cal) are an older unit of energy. Calories are bigger than kilojoules.

$$1 \text{ Cal} = 4.2 \text{ kJ}$$

 ISBN 9780170412650

EXAMPLE 2

A 500 mL chai tea latte with full cream milk contains 240 calories.
How many kilojoules is this?

Solution

Kilojoules are smaller than calories. To change to a smaller unit, multiply by the conversion factor.

$$240 \text{ Cal} = 240 \times 4.2 \text{ kJ}$$
$$= 1008 \text{ kJ}$$

Exercise 13.01 Burning energy

1 Mark is a 20-year-old bricklayer and has a very active lifestyle.

 Example 1

    a According to the table on the previous page, how many kilojoules does he require per day?

    b Mark eats a hearty breakfast containing 3250 kJ to give him energy for the day's work. What percentage (to the nearest whole number) of his daily energy requirements does Mark have for breakfast?

2 Stella is 36 years old. During the week she is inactive and on the weekend she is very active.

    a How much energy does she need each day during the week?

    b How much more energy does she require per day on the weekend than during the week?

3 Courtney, aged 18, is very thin and refuses to eat more than 6000 kJ per day. Every afternoon she works out at the gym for 2 hours. Courtney's mother is worried and knows that the type of gym workout Courtney does burns 2500 kJ/hour.

    a How much energy does Courtney use at the gym each day?

    b How much energy does Courtney have left from her diet for the remaining 22 hours in the day?

4 Use the conversion 1 Cal = 4.2 kJ to copy and complete each statement. Answer to the nearest whole number.

 Example 2

    a 500 Cal = ___ kJ   **b** 360 kJ = ____ Cal   **c** 68 Cal = ___ kJ

    d 25 kJ = ____ Cal   **e** 2460 kJ = ___ Cal   **f** 10 800 kJ = ___ Cal

5 **a** The treadmill at the gym shows that Samantha has used 550 Cal. How many kilojoules is this?

    b Samantha wants to use 2500 kJ on the machine. How many more calories does she have to burn?

6 This graph shows children's daily energy requirements.

a Do boys or girls need more energy per day?

b At what age do boys and girls need approximately the same amount of energy per day?

c How many more kJ/day do 16-year-old boys need than 16-year-old girls?

d At what age do boys require approximately 7500 kJ/day?

e Calculate the weekly energy requirements for a 12-year-old girl.

f Jon is an 8-year-old boy who usually eats 8600 kJ of food per day. Is he eating an appropriate amount?

g If Jon continues to eat 8600 kJ/day, will Jon's weight increase or decrease?

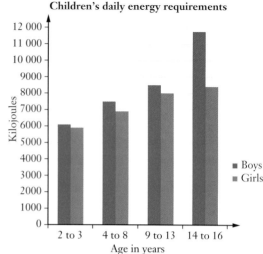

Children's daily energy requirements

7 This table shows the average length of time it takes an 18-year-old to burn up 1000 kJ for different activities.

Activity	Time required to use 1000 kJ
Sleeping	4 hours
Eating	3 hours
Working in class Studying Watching TV	2 hours 30 minutes
Walking	1 hour
Bike riding	50 minutes
Swimming	30 minutes

a How much energy does Scott use swimming for 1 hour?

b How long will it take Jessica to use 500 kJ while she is sleeping?

c Suzie, aged 18, leads a very active life. This pie chart shows the number of hours each day that she spends on different activities.

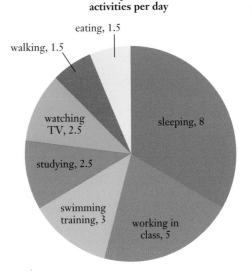

Hours Suzie spends on different activities per day

eating, 1.5
walking, 1.5
sleeping, 8
watching TV, 2.5
studying, 2.5
swimming training, 3
working in class, 5

   i According to the table on page 312, how much energy does Suzie need per day?

   ii Use the chart and the table on the previous page to calculate the amount of energy Suzie uses per day.

   iii If Suzie eats food that provides the energy requirement given in part **i**, is it enough for what she uses per day?

   iv What recommendations would you give Suzie concerning her lifestyle?

13.02 Energy in food

The food we eat supplies us with energy. This table shows the amount of energy stored in some normal-sized serves of food.

Investigating fast food

Food	kJ	Food	kJ	Food	kJ
Roast lamb, gravy	1064	Apple pie	1380	Apple	243
Potato bake	1175	One slice of buttered bread	520	Large tomato	120
				Small tomato	30
Mixed grill	2600	Yoghurt	315	Chocolate biscuit	493
Steak	3900	Muesli	1470	Can of soft drink	372
Bacon (2 slices)	640	Eggs (2)	735	Orange juice	206
Fish	340	Cheese	115	Milk (glass)	628
Grilled chicken breast	1264	Ham (2 slices)	224	Milk for cereal	400
Chips	1425	Mixed nuts (100 g)	2640	Coffee with milk and sugar	295
Ice cream	810	Broccoli	98	Banana	546
Sauce for steak or chicken	246	Sauce for fish	265	Beef sausage	176

EXAMPLE 3

On a plane, Lucas was served a meal of grilled chicken breast with sauce, chips and broccoli. How many kilojoules were in this meal?

Imagefolk/Sylvain Grandadam

Solution

Use the table to look up the kJ content of each food item.

Chicken breast: 1264

Sauce: 246

Chips: 1425

Broccoli: 98

Add them.

Total kJ = 1264 + 246 + 1425 + 98

= 3033

Write your answer.

The meal contained 3033 kJ.

EXAMPLE 4

Marna had 200 g of mixed nuts. How long will it take her to use the energy from them when she is working out at the gym at the rate of 30 kJ/minute?

Solution

200 g of nuts = 2 serves (from the table).

Total energy = 2640 × 2

= 5280 kJ

Divide the number of kJ by the rate to find the time.

Time = 5280 ÷ 30

= 176 minutes

= 2 hours 56 minutes

Write your answer.

Marna will work out for 2 h 56 min to use the energy from the mixed nuts.

EXAMPLE 5

Athletes running at 15 km/h use 75 kJ/h per kg of their body weight. Claire's body weight is 58 kg. How much energy does she use when she runs for 30 minutes at 15 km/h?

Solution

Claire burns 75 kJ per hour per kg of her body weight. She weighs 58 kg, so each hour she burns 75×58.

Energy burned per hour $= 75 \text{ kJ} \times 58$
$$= 4350 \text{ kJ}$$

30 minutes is half of an hour. Divide by 2.

Energy burned in 30 minutes $= 4350 \div 2$
$$= 2175 \text{ kJ}$$

Write your answer.

Claire burns 2175 kJ when she runs for 30 minutes.

Exercise 13.02 Energy in food

1 Calculate the energy content of each meal.

Example 3

  a Dani's breakfast contains a serving of muesli with milk, an apple and a banana.

  b Felix's breakfast comprises 2 slices of ham, 2 beef sausages, one egg and one slice of buttered bread.

  c Milan ate a steak with sauce, a large tomato and a serving of chips for his lunch.

  d Emma had some grilled fish with sauce and a large sliced tomato for her lunch.

2 George is 16 years old. The menu shows what he eats in a typical day.

a Calculate the total number of kilojoules George eats in a typical day.

b An average 16-year-old boy should eat 11 800 kJ per day. What effect will George's typical day's food have on his body weight?

3 Amanda requires a daily diet of 8800 kJ. Use the food items in the table on page 315 to plan a healthy day's menu for Amanda.

4 A large chocolate cake contains 27 600 kJ. Suyen cut the cake into 8 equal slices.

a How many kilojoules are in each slice of cake?

b How many minutes of gym work, at 30 kJ/minute, will it take to use the energy in one slice of the cake?

5 This table shows the energy an average person uses per minute for different activities. Calculate the amount of energy an average person uses when:

a gardening for 30 minutes

b ironing for 45 minutes

c bricklaying for an hour

d circuit training for $1\frac{1}{4}$ hours

e sleeping for 8 hours

f walking for 50 minutes every day for a week

Activity	kJ/minute
Sleeping	4
Cleaning	15
Ironing	17
Bricklaying	17
Playing tennis	31
Gardening	23
Circuit training	53
Walking	23

6 The more you weigh, the more kilojoules you will burn doing different activities. The table shows the amount of energy per kg of body weight that teens and adults use in one hour in various activities.

Example 5

a Lisa weighs 45 kg. How much energy does she use during a 1-hour maths lesson?

b Ali's body mass is 75 kg. How many more kilojoules does he use in 8 hours sleep than Lisa?

c Marko started walking every day to try to lose weight. He currently weighs 128 kg. How much energy does he use when he walks for 30 minutes?

Activity	kJ used per hour per kg of weight
Sleeping	3.7
Driving a car	6.4
Schoolwork	6.8
Walking	18
Skiing (15 km/h)	43
Swimming (45-second laps)	29

d Filomena is a keen skier who weighs 63 kg. One Saturday she drove her car for 3 hours to the snowfields. Then she skied at an average speed of 15 km/h for 5 hours. How many kJ of energy did she use in the 8 hours?

e Every morning Scott spends an hour swimming 45-second laps of the pool as part of his training routine. During the hour, Scott uses approximately 2200 kJ. What is Scott's approximate body weight?

PS

Chapter problem

You've covered the skills required to solve the chapter problem. Can you solve it now?

MY DAILY ENERGY REQUIREMENTS

What you have to do

- Keep an activities/exercise diary for 1 week. Record the number of hours you spend on different activities each day. Classify the activities as sleeping, eating, schoolwork or study, low-level activities like watching TV, medium-level activities like walking and high-level activities like swimming, running or going to the gym.

- Also keep a food diary to record the number of kilojoules you eat each day. Remember to include 'hidden' kilojoules like those in tea/coffee, biscuits and other snacks.

- Use the results in your diary to calculate the kilojoules you need, on average per day. If you don't know the amount of kilojoules involved in any of your activities, use the Internet to find out.

- Do your kilojoules balance? What changes might you need to make?

13.03 Energy in electricity

Household
energy costs

Power
problems

Energy
consumption

Units of electricity

- A **watt** (abbreviation **W**) is a unit of electrical **power**.

- 1 kilowatt (**kW**) = 1000 watts.

- Electrical energy is measured in **kilowatt-hours (kWh)**, which is the amount of electricity used by a 1000 W load drawing power for one hour.

The running costs of different appliances can be calculated if you know the power rating or wattage of the appliance, your usage time and the electricity tariff (costs) for your area.

Shutterstock.com/Ksander

EXAMPLE 6

Lina has a 200 W TV that she uses for 6 hours per day. The domestic tariff for electricity is 34 cents per kilowatt-hour, written as 34 c/kWh.

a How many kWh of electricity will Lina's TV use each day?

b How much will the electricity cost?

Solution

a First change 200 W to kW by dividing
by 1000.

$$200\text{ W} = 200 \div 1000$$
$$= 0.2\text{ kW}$$

Find the electricity usage by multiplying by
the number of hours.

$$\text{Electricity usage} = 0.2\text{ kW} \times 6\text{ h}$$
$$= 1.2\text{ kWh}$$

b Find the electricity cost by multiplying by
the rate.

$$\text{Electricity cost} = 1.2\text{ kWh} \times 34\text{c}$$
$$= 40.8\text{ c}$$

Exercise 13.03 Energy in electricity

1 Copy and complete this table to calculate the typical daily and monthly costs of some home appliances. Assume electricity costs are 34 cents per kWh and that there are 30 days in a month.

Example 6

	Appliance	Power rating (W)	Hours used per day	kWh per day	Daily cost (cents)	Monthly cost ($)
a	Fridge (600 L)	800	24			
b	Clothes dryer	2400	1.5			
c	Washing machine	900	0.5			
d	Bathroom/fan/heater/light	1100	1			
e	Normal light globe	100	6			
f	Food processor	380	0.5			
g	Electric kettle	1500	1			
h	Hotplate (max setting)	1500	0.25			
i	Dishwasher	1900	1			
j	Toaster	650	0.5			
k	Vacuum cleaner	950	0.25			
l	Stereo system	40	5			
m	Hair dryer	1400	0.5			
n	TV (med to large)	550	6			
o	Iron	950	0.75			

2 Which of the appliances from question **1** are in your household? How many are there of each appliance?

3 Use the information from question **1** to calculate the approximate cost of electricity for running these appliances in your household for one year.

4 How close is your calculation here to your household's electricity bill? Give reasons why your calculation might be different to your actual bill.

Energy ratings

INVESTIGATION

THE ENERGY RATING SYSTEM

Some household appliances now display **energy rating labels** to help people choose energy-efficient appliances. These labels contain a star rating to show how efficient the appliance is – the more stars, the more efficient the energy use. The label also gives an estimate of the energy the appliance will use over a year.

Use the Internet to research the answers to these questions.

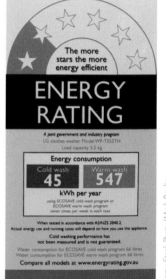

1 What information does the energy rating label on electrical appliances provide?

2 What does it mean for an appliance to have more stars than another?

3 Apart from the number of stars, what other important information is printed on the label?

4 There are 8 appliances that are required to carry energy rating labels. List them.

5 The 2 key features of the label are:

    a the star rating band: how many stars can this have and what does the coloured section indicate?

    b the energy consumption box: what does the number in this box indicate?

6 For all star-rated appliances other than air conditioners, the energy consumption figure is given in what units and for what period?

7 How is the information given for air conditioners?

8 Write down how the energy consumption figure can be used to calculate the annual electricity cost of the appliance.

9 Calculate the annual cost of running a refrigerator rated at 750 kWh/year if the electricity tariff is 34 cents/kWh.

13.04 Renee's gym training

Heart rates

The heart is a large muscle that pumps blood around our body. If it stops beating for more than a few minutes, we die. Our hearts beat faster during exercise. After exercise, the time it takes for our heart to return to its resting rate is a measure of our fitness.

Our **heart rate** or **pulse** is measured in beats per minute. The average rate is 72 beats/min.

We can use a simple formula to convert a pulse rate for any time period into a rate per minute.

Pulse

$$\text{Pulse in beats per minute} = \frac{\text{number of beats}}{\text{time in seconds}} \times 60$$

EXAMPLE 7

Darren takes his pulse during an exercise session and counted 52 beats in 20 seconds. Calculate his pulse in beats per minute.

Solution

$$\text{Pulse in beats per minute} = \frac{\text{number of beats}}{\text{time in seconds}} \times 60 \qquad \text{Pulse} = \frac{52}{20} \times 60$$

$$= 156 \text{ beats/min}$$

Exercise machines can generate electricity

Gyms can collect the energy people use on exercise machines and convert it to electricity. One person doesn't generate a lot of energy, but when the gym is full, the electricity generated can power the gym's TV and sound systems, which saves money.

Shutterstock.com/Lucky Business

EXAMPLE 8

At the gym, each exercise bike rider generates 100 watts of electricity per hour. On average, an exercise bike is used for 6 hours per day, 364 days per year (closed Christmas Day) and the gym pays 54c per kilowatt for electricity.

How much in electricity costs does each bike save the gym annually?

Solution

Calculate the electricity a bike generates each day by multiplying the rate by 6 hours.	Electricity generated per day $= 100 \times 6$ $= 600$ W
Multiply the daily amount by 364 to get the annual amount.	Electricity generated per year $= 600 \times 364$ $= 218\,400$ W
Divide by 1000 to change it into kilowatts.	$218\,400 \div 1000 = 218.4$ kW
Calculate the electricity cost by multiplying by the cost rate. Write 54c as $0.54.	Cost $= 218.4 \times \$0.54$ $= \$117.94$
Write the answer.	On average, each exercise bike saves the gym $117.94 in electricity costs per year.

Exercise 13.04 Renee's gym training

1 Renee's 3-month gym membership costs $234 and gives her 13 weeks unlimited use of the gym.

   a How much does Renee's membership cost per week?

   b Casual users of the gym are charged $11 per visit. Renee attends the gym 4 times per week. How much will her 3-month membership save her compared to the casual rate?

   c How many visits to the gym does a member need to make in 3 months to make membership worthwhile?

2 A gym is advertising special membership deals to attract new members.

   a Calculate the charge per week for each of the two plans.

   b If you were joining the gym, which plan would you choose? Why?

Special membership **deals**

$234
13 week membership
(Paid fortnightly)

$832
12 month (52 week) membership
(Paid as a lump sum at the beginning of each year)

3 Convert each pulse rate to beats per minute.

   a   25 beats in 20 s   **b**   16 beats in 15 s   **c**   65 beats in 30 s

Example 7

4 The gym manager is making a sign to help visitors convert their pulse for 15 seconds to beats/minute. Complete the missing values in the sign.

Heart beats in 15 seconds	Pulse in beats per minute
20	a
25	b
30	c
35	d
40	e
45	f
50	g

5 During her first aerobics class, Renee counted her pulse as 43 beats in 15 seconds.

   a   What is Renee's pulse in beats per minute?

   b   The best pulse rate during exercise is 60% of the maximum pulse rate. Renee is 16 years old. Use the sign to calculate her best exercise pulse rate.

   c   What advice would you give Renee if she exercises at an intensity that gives her a pulse of 43 beats in 15 seconds?

Maximum pulse rates in beats per minute

15 – 19 years	180
20 – 29 years	170
30 – 39 years	160
40 – 49 years	150
50+ years	140

6 Renee's resting breathing rate (respiration rate) is 16 breaths/minute. When she runs on the mini-trampoline, she likes to run at a pace that gives her a 75% increase in her respiration rate. What respiration rate does Renee aim for when she uses a mini-trampoline?

PS

7 As a warm-up exercise, Renee rode an exercise bike at a speed of 20 km/h for 5 minutes.

   a   What fraction of one hour is 5 minutes?

   b   Use your answer to part **a** to calculate how many kilometres (correct to 2 decimal places) Renee rode on the bike during the 5 minutes.

8 In a circuit class, participants change fitness machines every 30 seconds. There are 28 different activities in the circuit. These are completed at workstations.

   a   How long will it take Renee to complete one circuit using each workstation once?

   b   In a circuit class the workstations are used for 25 minutes. How many workstations will Renee use per class?

9 When Renee is using the fitness machines, she generates 85 watts of electricity per hour. She uses the machines for 3 hours per week.

    a How many watts of electricity does Renee generate on the machines per week?

    b Calculate the amount of electricity Renee will generate during her 13-week membership. Express your answer in kilowatts.

    c Calculate the value of the electricity Renee generates in 13 weeks at 54c/kilowatt.

10 The gym has 24 treadmills. On average, each treadmill is in use for 10 hours per day, 364 days per year. Members using a treadmill generate 110 watts of electricity per hour.

    a Calculate the total number of kilowatts of electricity generated by members on the treadmills per day.

    b How many kilowatts of electricity do members generate on the treadmill per year?

    c At 54c/kilowatt, calculate the total annual value of the electricity generated on the treadmills.

KEYWORD ACTIVITY

Match the terms in the left column with their meaning on the right.

Term		**Meaning**
1 calorie	**a**	A word used in rates that means 'for each'.
2 kilowatt-hour	**b**	A unit for measuring electrical power, equal to 1000 watts.
3 kilojoule		
4 kilowatt	**c**	The rate measured in beats/minute.
5 speed	**d**	An old unit for measuring the energy in food.
6 per	**e**	A measurement involving 2 different types of units.
7 pulse		
8 rate	**f**	The metric unit for measuring energy in food.
	g	A rate that compares distance travelled with the time taken to travel that distance.
	h	A unit for measuring electrical energy.

ISBN 9780170412650

SOLUTION ^{TO} _{THE} CHAPTER PROBLEM

Problem

For morning tea, Renee had a chocolate muffin (825 kJ) and some coffee with milk and sugar (295 kJ). Afterwards, Renee goes for a walk, using 23 kJ of energy per minute. How long will Renee have to walk to use all the energy she consumed at morning tea?

Solution

WHAT?

STAGE 1: WHAT IS THE PROBLEM? WHAT DO WE KNOW?

Calculate the time Renee has to walk to use the energy from morning tea. She had a muffin (825 kJ) and coffee (295 kJ).

SOLVE

STAGE 2: SOLVE THE PROBLEM

Total energy Renee consumed = 825 + 295

$\qquad\qquad\qquad\qquad = 1120$ kJ

Number of minutes walking $\quad = 1120 \div 23$

$\qquad\qquad\qquad\qquad\qquad = 48.7$ minutes

$\qquad\qquad\qquad\qquad\qquad \approx 50$ minutes

CHECK

STAGE 3: CHECK THE SOLUTION

50 minutes is reasonable for burning over 1000 kJ of energy at 23 kJ/minute.

The time is in a reasonable range.

PRESENT

STAGE 4: PRESENT THE SOLUTION

Renee should walk for about 50 minutes to use up the energy she had at morning tea.

13. CHAPTER REVIEW

Healthy figures

1 Use the conversion 1 Cal = 4.2 kJ to complete each statement.

    a 250 Cal = _____ kJ   **b** 5754 kJ = _____ Cal

2 When riding a bike, Alex takes 50 minutes to burn 1000 kJ.

    a How many kJ does Alex use riding a bike for 25 minutes?

    b How long does it take Alex to use 3500 kJ when bike riding?

3 Kelly is an active pregnant woman aged 24.

    a How many more kilojoules does she require per day when she is pregnant than when she is not?

    b Use the table on page 315 to suggest foods Kelly could add to her diet to make up the kilojoules she requires while she is pregnant.

4 Marc ate a mixed grill and apple pie with ice cream for his dinner.
Use the table on page 315 to calculate the number of kilojoules in Marc's dinner.

5 Adults running at 15 km/h use 75 kJ/h per kilogram of their body weight. Kristy's body weight is 64 kg. How many kilojoules does she use when she runs for 45 minutes at 15 km/h?

Shutterstock.com/lzf

---

6 How much electricity is used (in kWh) when a 100 W light globe is on for 5 hours?

Exercise
13.03

7 Tamara takes about $1\frac{1}{2}$ hours per week to iron her clothes. If she uses a 200 W iron and the cost of electricity is 28 c/kWh, calculate the cost of using the iron each week.

Exercise
13.04

8 Lenny's heart beats 11 times in 10 seconds. Calculate his pulse in beats per minute.

Exercise
13.04

9 Marcus generates 96 watts of electrical energy per hour when he uses an exercise bike. Calculate the total value of the energy he generates using the bike for 45 minutes/day for 6 days. Use 55c/kilowatt-hour as the value of the electricity.

Exercise
13.04

14.

PAYING TAX

Chapter problem

Hamish is preparing his income tax return. The annual payment summary shown below is for his day job only. Hamish also earned $3640 and paid $1040 tax in his part-time job. He has $460 in allowable tax deductions.

Will Hamish receive a tax refund?

PAYG payment summary		
Employee	Hamish Donald	
Employment period	1 July 2018–30 June 2019	
	Total income	$48 230
	Total tax withheld	$8461

14.01 Gross and net pay
14.02 Taxable income and tax refunds
14.03 Income tax and Medicare levy
14.04 Budgeting
Keyword activity
Solution to the chapter problem
Chapter review

WHAT WILL WE DO IN THIS CHAPTER?

- Calculate net pay after making deductions from gross pay
- Calculate taxable income after making allowable deductions from income
- Calculate income tax and Medicare levy
- Calculate tax payable or refund owing after PAYG tax has been paid
- Prepare a household budget

HOW ARE WE EVER GOING TO USE THIS?

- Checking the amount of tax deducted from our pay
- Making sure we don't pay too much tax
- Keeping accurate records and collecting receipts to calculate allowable tax deductions
- Calculating whether we're entitled to a tax refund
- Using a budget to manage our income and spending

14.01 Gross and net pay

Everyone who's had a job knows that the amount of money they've earned isn't what they end up getting. Most workers pay income tax under the **PAYG** (Pay As You Go) system, where your employer estimates and takes out the tax from your **gross pay** before they give you your **net pay**. Some people have other amounts deducted from their pay, such as health insurance payments, superannuation contributions or union fees. What's left of their pay after tax and other **deductions** is called their 'take-home pay'.

Gross and net pay

Gross pay is the total amount earned.

Net pay = gross pay − tax

Take-home pay = gross pay − tax − other deductions

EXAMPLE 1

Chris earns $155 per shift in his part-time job stacking supermarket shelves late at night. He works 3 shifts per week and each fortnight his employer deducts $45.60 for PAYG tax.

a   What is Chris' fortnightly gross pay?

b   Calculate Chris' fortnightly net pay.

Solution

a   Chris works 3 × 2 = 6 shifts per fortnight.

$$\text{Gross pay} = 6 \times \$155$$
$$= \$930$$

b   Net pay = gross pay − tax.

$$\text{Net pay} = \$930 - \$45.60$$
$$= \$884.40$$

Exercise 14.01 Gross and net pay

Example
1

1   Dylan earns $680 per week for working 4 days per week and his employer deducts $84.50 for PAYG tax.

    a   Calculate Dylan's net weekly pay.

    b   How much PAYG tax does Dylan pay per year?

2 Complete the missing values in the table.

Gross weekly income	Weekly tax	Net weekly income
$720	$32	**a**
$860	**b**	$822
c	$75	$1100
$1258	$137	**d**
$1450	**e**	$1139

3 Sarah earns $17.65 per hour for a 35-hour week. Each week, she has $54 deducted from her pay for PAYG tax and $11 for health insurance.

   a How much does Sarah earn per week?

   b Calculate Sarah's weekly take-home pay.

4 Grant's annual tax bill is $9704. How much tax should he pay per week to cover his annual bill?

5 Aisha's gross salary is $80 000 and she calculated that her annual tax bill is $17 547. Aisha is paid fortnightly.

   a How much PAYG tax should she pay each fortnight?

   b Calculate her fortnightly net pay.

6 Juan's PAYG payment summary from his employer showed that he paid $5845 in PAYG tax last year. His gross salary was $35 700.

   a How much PAYG tax was deducted from Juan's pay each week?

   b What was his net weekly pay?

7 Xander is constructing a spreadsheet to calculate annual net pay.

	A	B	C	D	E
1	Net annual Income				
2	Enter gross fortnightly pay in D2				
3	Enter fortnightly PAYG tax in D3				
4		Annual net pay			
5					
6					

What formula will he need for cell D4?

8 Boun is a junior barrister in a city law firm. His gross fortnightly pay is $3692 and his net fortnightly pay is $2656. How much tax is deducted from Boun's pay each year?

14.02 Taxable income and tax refunds

Running a country is an expensive business. The Australian Government provides the community with public healthcare, social services, border security, education, highways and much more. **Income tax** is one of the taxes the government uses to pay for these things.

PAYG tax is deducted from your pay and is only an estimate of the income tax you need to pay. The actual amount of tax that you're supposed to pay is calculated at the end of the financial year (30th June) and is based on your total income over the previous 12 months.

After 30th June, you are required to complete a **tax return** to tell the Australian Tax Office (ATO) how much money you've earned and how much PAYG tax you've paid. The ATO then calculates the actual income tax payable and decides whether you've paid too much tax or not enough. If you've paid too much, you will receive a **tax refund**. If you haven't paid enough, you will be sent a bill to pay the **tax debt**.

If you donate some of your income to charities or spend it on work-related items and activities, then you do not have to pay tax on those amounts. These amounts are 'tax deductible' and are called **allowable tax deductions**. We deduct these amounts from our annual income and then income tax is calculated on the remaining amount, called our **taxable income**.

> Taxable income = total income – allowable tax deductions

EXAMPLE 2

Last financial year, Rowan's gross salary was $65 000 and he earned $420 in interest from his bank account. He had allowable tax deductions of his $290 union membership fees and an $80 donation to the RSPCA. How much is Rowan's taxable income?

Solution

Total income = gross salary + bank interest.	Total income = $65 000 + $420
	= $65 420
Find total tax deductions.	Deductions = $290 + $80
	= $370
Taxable income = total income – deductions.	Taxable income = $65 420 – $370
	= $65 050

ISBN 9780170412650

EXAMPLE 3

Last financial year, Rowan paid $302 each week in PAYG tax instalments. The tax office calculates that on an income of $65 000 his income tax should be $15 095. Will Rowan receive a tax refund or a tax debt? How much?

Solution

Calculate the total PAYG tax paid.	PAYG tax paid = $302 × 52 = $15 704
Compare this to the amount Rowan has to pay ($15 095).	Rowan paid more than was required ($15 704 > $15 095) so he will receive a tax refund.
Calculate the difference to find the refund.	Tax refund = $15 704 − $15 095 = $609 Rowan will receive a $609 tax refund.

Exercise 14.02 Taxable income and tax refunds

1 Nabil's gross annual pay is $69 600 and he earns $1280 in bank interest. He has the following allowable tax deductions: trade tools $370, mobile phone $413, home office equipment $177.

    a Calculate Nabil's gross annual income.

    b How much is Nabil's taxable income?

Example 2

2 Last year, Sandy paid $11 865 in PAYG tax. The tax office informed Sandy that her income tax payable is $11 450. How much will Sandy receive as a tax refund?

3 Madeleine's weekly net pay is $890 and her weekly gross pay is $995.

    a How much does Madeleine pay in PAYG tax each week?

    b Madeleine's income tax payable is $4790.50. Will she receive a tax refund or a tax bill? What amount is involved?

4 Every fortnight, Sunny earns $1775 in net pay after paying $260 in PAYG tax.

    a Calculate his gross pay.

    b What is Sunny's annual gross pay?

    c Sunny's income tax payable is $7897.75. Will he receive a refund or a bill? Justify your answer.

5 Abdul earns $406 per week. During the year, he paid $160 in union fees and $65 in work-related expenses.

    a Calculate his gross annual income.

    b What is Abdul's annual taxable income?

    c Each week, Abdul paid $47 in PAYG tax. How much tax did Abdul pay during the year?

    d The tax office calculated that Abdul's income tax was $2250. Will Abdul receive a refund or will he have to pay more tax? How much?

6 Mrs Durham pays $178.20 PAYG tax every week. At the end of the year, the ATO calculated that the actual tax she was required to pay was $9320.40. Did she receive a refund or did she have to pay more tax?

7 Peter's salary was $77 500 and he earned $165.20 from his shares. His allowable taxation deductions were conference fees of $310 and a donation of $60 to the Salvation Army. How much was his taxable income?

8 Each week, Gianna earned $1635 and paid $496.18 in PAYG tax. Her work-related travelling expenses of $199 were tax deductible.

    a Calculate Gianna's gross annual income.

    b How much PAYG tax did she pay for the year?

    c How much was her annual taxable income?

    d The ATO calculated Gianna's tax to be $19 113.84. Did she receive a refund? Explain your answer.

14.03 Income tax and Medicare levy

This table shows the tax rates for different income brackets.

Taxable income	Tax on this income
0 – $18 200	Nil
$18 201 – $37 000	19c for each $1 over $18 200
$37 001 – $87 000	$3572 plus 32.5c for each $1 over $37 000
$87 001 – $180 000	$19 822 plus 37c for each $1 over $87 000
$180 001 and over	$54 232 plus 45c for each $1 over $180 000

© Australian Taxation Office for the Commonwealth of Australia

There is also a tax that goes towards funding Medicare, the public health system, called the **Medicare levy**.

For most people, the Medicare levy is 2% of their taxable income.

Note: The Medicare levy is due to increase to 2.5% in July 2019.

EXAMPLE 4

Rodney's gross salary is $46 248 and he has allowable tax deductions totalling $7310. Calculate Rodney's income tax and Medicare levy.

Solution

Calculate Rodney's taxable income.	Taxable income = $46 248 − $7310
	= $38 938
$38 938 is in the $37 001 to $87 000 row of the tax table.	Income tax = $3572 + 0.325 × ($38 938 − $37 000)
$3572 plus 32.5c for each $1 over $37 000 (which means 32.5% or 0.325).	= $4201.85
Medicare levy is 2% of taxable income.	Medicare levy = 0.02 × $38 938
	= $778.76

The government uses **tax file numbers** (TFN) to record what we earn from our jobs, shares and investments. If we don't provide our TFN to the government, then we must pay PAYG tax at the maximum income tax rate (45%) every payday, before claiming it back at the end of the financial year. The government does this to minimise tax avoidance.

Exercise 14.03 Income tax and Medicare levy

1 Calculate the income tax, Medicare levy and the total tax for each taxable income.

Example 4

   a $45 200

   b $36 960

   c $49 625

   d $122 500

   e $186 000

2 Marianne's gross salary is $46 230 and her total allowable deductions are $7520.

   a What is her taxable income?

   b Calculate her income tax.

   c How much is Marianne's Medicare levy?

   d Marianne's PAYG summary from her employer shows that she has paid $4520 in PAYG tax. Will this cover her income tax and Medicare levy?

   e How much extra will she have to pay?

3 Muspah's gross salary is $126 450 and he has $11 200 of allowable deductions.

  a How much income tax is he required to pay?

  b Because Muspah has a high taxable income and he doesn't have any private health insurance he has to pay a Medicare levy surcharge. His Medicare levy is increased to 3.25% of his taxable income. Calculate Muspah's Medicare levy, including the surcharge.

  c How much extra does Muspah have to pay for Medicare because he doesn't have private health insurance?

4 Kait works in a hospital. She recently received a $1600 pay rise that took her taxable income from $75 200 to $76 800.

  a How much will her income tax increase as a result of her pay rise?

  b How much will her Medicare levy increase as a result of her pay rise?

5 Jacob's taxable income is $50 000 and his friend Garth's taxable income is $150 000. Garth claimed that because his income is three times Paul's income he pays three times as much tax as Paul. Use calculations to determine whether Garth's statement is correct.

6 Last financial year, Tina earned $56 800 from her full-time building job and $26 000 from her weekend work. Her allowable deductions for the year totalled $5600.

  a Calculate Tina's income tax and Medicare levy.

  b During the year, Tina's full-time employer deducted $10 300 in PAYG tax. How much does Tina owe the tax office?

7 Single people with a taxable income of $90 000 and no private health insurance are required to pay the Medicare levy surcharge of 1% of their taxable income.

  a Calculate the Medicare levy plus surcharge; that is, a 3% Medicare levy on a taxable income of $90 000.

  b How much extra is this compared to a similar person who *does* have private health insurance?

  c 'Basic plus' private health insurance for young singles costs $16.05 per week. How much less than the Medicare levy surcharge does this health insurance cost?

8 This year, Jules' Medicare levy was $1285, which was 2% of his taxable income. Calculate Jules' taxable income.

> **Chapter problem**
>
> You've covered the skills required to solve the chapter problem. Can you solve it now?

COMPLETING A TAX RETURN

In Chapter 10, you looked at your possible career in the 'My future career' investigation. Imagine you have just completed a financial year working in one of the careers recommended for you.

Part 1: Your financial details

1 Enter your 9-digit tax file number if you have one or make one up if you haven't.

2 Research the income you can expect to earn per year in your recommended career.

3 Calculate 20% of the value of your income and assume you have paid that amount in PAYG tax.

4 List any expenses you could incur that are work-related.

5 Decide whether you will have private medical insurance and research the annual insurance cost.

6 Record any bank interest or dividends from shares that you think you might receive.

Part 2: Completing the tax return form

1 Visit the Australian Taxation Office website, www.ato.gov.au, and download a tax return form for individuals.

Australian
Tax Office

2 Print the form, then open the 'Individual tax return' instructions document. Printing this document is not required.

3 Enter the information from Part 1 into the tax return form. If you don't know how to complete any section, ask your teacher for help or look at the relevant pages of the tax return instructions.

4 Determine your taxable income, the income tax you are required to pay, and your Medicare levy.

5 Determine the amount of your tax refund or any extra tax you have to pay.

6 You can use the tax calculators on the ATO website to check your calculations.

When you have a real job and tax file number, you can complete your tax return online.

Budgeting
scenarios

Budget grid

14.04 Budgeting

Have you ever wondered what happens to your money? It's a good idea to have a plan so that you don't waste it. A **budget** lists your expected income and expenses, and can help you to manage your money.

Income covers all the money you might earn.

Expenses covers all the ways you might spend that money, and there are two types:

• **Fixed expenses** are costs that are essential and must be paid. Some are the same amount each time, such as rent. Others aren't always the same, such as food.

• **Discretionary expenses** are amounts that you often spend but which aren't essential, such as entertainment or magazines.

A budget needs to balance your income and expenses so that you have enough money for everything you require and some left over to save for special items, such as a car or holiday.

Shutterstock.com/wavebreakmedia

EXAMPLE 5

Ashleigh works part-time while studying. She receives an allowance from her parents of $100 per week and she earns $120 from her job. She pays $80 per week in rent and spends $30 per week on food. She averages $10 per week for her phone and $20 per week on clothes. She divides the remainder equally between entertainment and savings.

a   Create a budget for Ashleigh for a week.

b   Ashleigh's rent is increased by $25 per week. How would she need to adjust her budget for this increased expense?

Solution

a List income and expenses and make sure total expenses equals total income.

Income		Expenses	
Allowance	$100	Rent	$80
Earnings	$120	Food	$30
		Phone	$10
		Clothes	$20
		Entertainment	$40
		Savings	$40
Total income	$220	**Total expenses**	$220

Total of fixed expenses = $80 + $30 + $10 + $20 = $140

Remainder available for entertainment and savings = $220 − $140 = $80

Divided equally = $80 ÷ 2 = $40

b The $25 increase in rent means Ashleigh now has $25 less to spend on entertainment and savings. She has $80 − $25 = $55 to divide between entertainment and savings. She could still divide this amount equally between the two ($27.50 each) or she could spend less on entertainment to continue to save $40 per week.

Exercise 14.04 Budgeting

1 Lily owns a car and has the following expenses each year:
- registration $349
- Compulsory Third Party (CTP) insurance $795
- comprehensive insurance $1110
- maintenance bills of $790.

She spends $53 per week on petrol.

a How much does Lily spend on her car each year?

b How much should she set aside in her weekly budget to cover her car expenses?

2 Mitchell works in an office during the week and in the bar at the local club on weekends. He earns $620 per week from the office job and $215 from the club. He pays $280 per week in rent and spends an average of $60 per week on food. His phone costs him $20 per week and travel expenses are $60 per week. The remainder of his income has to be divided between entertainment, clothes and savings.

a Create a budget for Mitchell for one week.

b Mitchell is considering buying a car. He would no longer have public transport expenses but he would need to allow $100 per week to pay off a loan and $75 for car expenses. Create a new budget for Mitchell.

3 Marko is an apprentice mechanic. His take-home pay is $790 per week. This table shows his weekly expenses.

Item	Amount
Board	$120
Phone	$21
Clothes	$65
Car	$112
Entertainment	$72
Other expenses	$88
Savings	
Total	$790

a How much does Marko save each week?

b Calculate his net annual income.

c How much is Marko able to save each year?

4 This table shows Shania's monthly budget.

Income		Expenses	
Part-time job	$290	Clothes	$140
Babysitting	$130	School needs	$32
		Entertainment	$50
		Phone	$55
		Fares	$23

a Calculate Shania's monthly income and expenses.

b Calculate the amount she is able to save each year.

c Shania would like to increase her savings so that she can go on an end-of-year holiday. Suggest three ways she could do this.

5 Sanjeev has taken a second job to save for a new car. His budget for a week is shown below.

Income		Expenses	
Main job	$750	Rent	$225
Second job		Travel	$56
		Food	$117
		Clothes	$55
		Entertainment	$75
		Bills	$157
		Savings	
Total	$908	Total	$908

a Calculate how much Sanjeev earns from his second job.

b Calculate how much he can save each week.

c If the car costs $25 000, how long would it take Sanjeev to save this money?

d Suggest ways Sanjeev could save more per week so that he can buy his car sooner. Create a new budget for Sanjeev.

ISBN 9780170412650

MY BUDGET

In this investigation, you are going to prepare two budgets: one for a typical school leaver and one for yourself based on your own choice of job when you leave school.

Part A: Typical school leaver

INCOME:

My budget

- You have finished school.

- After school, you enrolled in a TAFE course for a year.

- You are now working and earn $27 040 p.a.

- You don't have a partner or children, but you need to move out of home.

- Your new home will be 4 km from where you work (fast 50-minute walk).

1 Download the 'My budget' worksheet from NelsonNet which contains the **Weekly tax tables, Budget guidelines** and the **Lifestyle costs** provided for this activity.

2 Copy and complete this budget form based on the information, or use the worksheet.

Weekly income		$
Gross weekly income	Equals yearly wage ÷ 52	
Deduct tax (weekly)	From table (tax-free threshold, no leave loading)	
Net weekly income		
Regular weekly expenses		**$**
Housing	Mortgage, rent or share	
Transport	Car running costs, registration, CTP insurance etc. or train, bike or walking	
Personal spending	Clothing	
	Hair, grooming, cosmetics etc.	
Food	Groceries, including pet food	
Utilities	Phone expenses	
	Electricity, gas	

Total regular expenses		$
Discretionary weekly expenses		$
Insurance	Home and/or contents	
Insurance	Car – comprehensive	
Insurance	Health	
Entertainment	Pay TV, books, magazines, music, movies etc.	
Recreation	Sport, holidays	
Technology	Internet, laptop	
Total discretionary expenses		$
Net weekly income		
Total weekly expenses (regular + discretionary)		
BALANCE		$

Part B: You in your preferred occupation

Now you will repeat the activity in Part A but to create a budget based on your chosen occupation.

- The aim of a good budget is not simply to maximise savings.

- It is more to do with providing cost amounts that reflect the sort of lifestyle you want and ensuring these are within the income you have at your disposal.

- Your budget costing needs to cover the costs of the essentials of living (e.g. regular expenses such as food, shelter).

- For discretionary expenses, weekly amounts should allow you to afford the extra activities you wish to pursue, whether it is entertainment, holidays, fashion, or a more expensive car.

- You also need to consider carefully, within your budget, the pros and cons of taking out insurance of various types and explain the decisions you make.

- Consider carefully the advantages and disadvantages of taking out different types of insurance, whether you can afford them and explain the decisions you make.

- Assume you are single and have bought or are renting your new home, located 4 km from where you work (fast 50-minute walk).

My Future

1 Write your preferred occupation and determine the gross wage for that occupation as follows.

- Go to the **My Future** website.

- Click on **Occupations**, then find your occupation.

- Select **Prospects**, then **Full time weekly earnings**.

- Write the **Weekly income** into your budget.

2 Find the tax to be deducted from gross pay, using ATO *weekly* tax tables. These can be found on the 'My budget' worksheet or downloaded from the Australian Tax Office website and searching for the table of 'Weekly withholding amounts'.

Australian Tax Office

3 Regular income and Weekly expenses (regular) *must* have an amount in the last column.

4 Weekly expenses (discretionary) can be left blank as a means of saving money, but you must explain and justify your decision.

5 For this exercise, you are *not* to make up your own figures; you *must* use the prices listed in the **Budget guidelines** and **Lifestyle costs**.

6 Some of the prices/costs are per week, others are per fortnight, per month, or per year. You must convert *all* amounts to a weekly cost.

For *every cost category* (both regular and discretionary), you will need to show:

• a short explanation of what choice you have made and why;

• the calculations/working you have done to arrive at the amount you have used.

Shutterstock.com/jesterpop

TAX CROSSWORD

Earning and
taxation
crossword

Copy this crossword and complete it using the clues given in the paragraphs below.
NelsonNet also contains a different crossword called 'Earning and taxation'.

The Government needs money to provide public facilities such as hospitals, **2**_____,
roads, public transport and **5**_____ services. Part of the funding for these facilities comes
from income tax. The **3**_____ **8**____ helps to pay for public health and medical services.

Income tax is charged on your taxable **6**_____ which is your total income less your
allowable **10**_____. If your taxable income is lower than the tax-**1 across**_____
threshold you don't have to pay any tax. Each **11**____, fortnight or month when
employees receive their pay, the tax has already been taken out as part of the
4_____ (PAYG) system.

At the end of the **1 down**_____ year, you complete a tax return to inform the ATO
about the total income you have earned and the **9**_____ you have paid. If you have paid
too much tax, you will receive a tax **7**_____, but if you haven't paid enough tax you will
receive a bill for more.

SOLUTION TO THE CHAPTER PROBLEM

Problem

Hamish is preparing his income tax return. The annual payment summary shown below is for his day job only. Hamish also earned $3640 and paid $1040 tax in his part-time job. He has $460 in allowable tax deductions.

Will Hamish receive a tax refund?

PAYG payment summary		
Employee	Hamish Donald	
Employment period	1 July 2018–30 June 2019	
	Total income	$48 230
	Total tax withheld	$8461

Solution

STAGE 1: WHAT IS THE PROBLEM? WHAT DO WE KNOW?

Has Hamish paid too much tax and is he entitled to a refund?

WHAT?

We know his income and tax paid from 2 jobs and we know his allowable tax deductions.

STAGE 2: SOLVE THE PROBLEM

Calculate Hamish's income tax first.

SOLVE

Hamish's total income = $48 230 + $3640 = $51 870

Hamish's taxable income = $51 870 – $460 = $51 410

According to the income tax table on page 336, the income tax payable on taxable incomes between $37 001 and $87 000 is:

$3572 plus 32.5c for each $1 over $37 000.

Income tax = $3572 + 0.325 × ($51 410 – $37 000)

= $8255.25

PAYG tax already paid = $8461 + $1040 = $9501

Hamish has paid too much tax.

Tax refund = $9501 – $8255.25 = $1245.75

CHECK

STAGE 3: CHECK THE SOLUTION

Hamish's taxable income was $51 410 and he paid $8461 + $1040 = $9501 PAYG tax, which is quite high for a low-income earner. It's reasonable he will receive a refund. Check the calculations.

PRESENT

STAGE 4: PRESENT THE SOLUTION

Hamish will receive a $1245.75 tax refund.

Paying tax

1 Ava earns $19.50 per hour for a 36-hour week. Each week, she has $84 deducted from her pay for PAYG tax and $14 for private health insurance.

   a How much does Ava earn per week?

   b Calculate Ava's weekly take-home pay.

2 Jim's salary was $87 500 and he received $2965 interest on his investment account. His allowable tax deductions were $340 for uniform and shoes, $460 for tools and a donation of $80 to cancer research. How much was his taxable income?

3 Pia's gross annual income is $65 850 and she has allowable deductions totalling $1250. Calculate her taxable income and Medicare levy.

4 Use the income tax table on page 336 to calculate each person's income tax.

   a Claire's taxable income is $82 500.

   b Will's annual taxable income is $26 500.

5 Every January, Julie has to register and insure her car. Each year, she has trouble paying these expenses because she is short of money after Christmas. This year, her January car registration and insurance costs totalled $2040. What do you recommend she do this year to be ready for these expenses next January?

6 Billy is going to move out of home and live in a shared, rented flat with friends. His net fortnightly pay is $780. This is the fortnightly budget he prepared:

Item	Budget
Car repayments	$385
Phone	$52
Entertainment	$100
Food	$160
Drink	$70
Savings	$40
Total	**$780**

   a Find at least 3 things that are wrong with Billy's budget.

   b What advice can you give Billy?

15.

THAT'S BIASED

Chapter problem

Juanita is researching political views in her community. One Wednesday morning she stands at an intersection on the main street of town. She asks people passing, 'Are you going to vote Liberal or Labor in the next election?'.

Will Juanita's data be reliable and unbiased? If not, how can she improve her methods?

WHAT WILL WE DO IN THIS CHAPTER?

- Design a questionnaire correctly and consider target populations
- Detect bias in our questionnaires, samples and statistical data
- Minimise bias in the statistical process

HOW ARE WE EVER GOING TO USE THIS?

- When creating questionnaires to collect data
- When we evaluate the results of surveys
- When we read or consider information reported in the media

15.01 Questionnaires

Using questionnaires is one of the most common ways to collect information.
A good questionnaire:

- uses simple language

- has clear and precise questions

- meets privacy requirements: unauthorised people cannot access your information

- is free from bias

Bias is an unwanted influence in sampling or questionnaires that favours a particular
section of the population unfairly. Bias produces unreliable results because the sample or
questionnaire answers are not truly representative of the population.

Exercise 15.01 Questionnaires

Andrew has designed the first draft of a questionnaire to survey club members about the
entertainment that the United Club provides, including trips away. Andrew's draft is given
below. Read the questionnaire and answer the questions that follow.

United Club   **Help us help you!**

Please answer this short questionnaire to help the United Club better serve you.

1 How old are you?   **2** Are you male or female?

☐ Less than 30 years   ☐ Male

☐ Between 31 and 40 years   ☐ Female

☐ Between 41 and 50 years

☐ Over 50 years

3 Which of the Club's entertainments do you attend?

4 How often do you attend the following Club activities?

Concerts   Films   Dance nights   Trips away

5 Do you think the entertainment offered is reasonably priced?

6 Should the club offer more types of entertainment and, if so, what should it offer?

7 Are there any other comments you would like to make about the Club's
entertainment program?

Thank you for taking the time to complete this questionnaire. Please hand it in at the bar.

NELSON QMATHS 11. Essential Mathematics   ISBN 9780170412650

1 **Closed questions** give a number of options for the answer. How many questions in the questionnaire are closed?

2 **Open-ended questions** provide no options and require the person to write their own answer. How many questions in the questionnaire are open-ended?

3 The first question about age has a problem in the options given. What error has Andrew made? (Hint: where would a person aged 30 tick?). Rewrite the options to this question to fix this error.

4 When Andrew asks 'Which of the Club's entertainments do you attend?' it is not clear exactly what he means. Club members could be unsure which of the Club's activities is classified as entertainment. Rewrite the question to make it clear.

5 When Andrew asks 'How often do you attend the following Club activities?' he will get a range of answers. Analysing the data will be easier if Andrew provides some options. Suggest a series of tick boxes that Andrew could provide for people to record their answers.

6 Would all Club members agree on what 'reasonably priced' means? Suggest how Andrew could change this question to make it clearly understood.

7 Write another question about the cost of entertainment for this questionnaire.

8 There is a risk that answers won't remain private if the questionnaires are handed in at the bar. Suggest a more secure way to collect the questionnaire forms.

9 Do you think Andrew has missed anything in his questionnaire? (Hint: how would people who don't attend the entertainment activities express their views?) Write 3 additional questions Andrew could use.

PS

Alamy Stock Photo/Directphoto Collection

15.02 Bias in questionnaires

The way a question is worded in a questionnaire can introduce bias. The wording of the questions can lead people towards a particular view. This can unfairly influence the way people answer the questions.

Andrew has written some questions to ask Club members about the club bistro.

a   How often do you eat at our fabulous bistro?

b   Rate your last meal: Great   Yummy   Quite nice

In what way are these questions biased? Rewrite them so they are not biased.

Solution

a   Using the word 'Fabulous' encourages the person answering the question to give a positive answer.

It would be better to ask: 'How often do you eat at the bistro?'

b   This question doesn't give any negative options. The question could be: 'Rate your last meal:

Poor   Average   Good'

Exercise 15.02 Bias in questionnaires

1 For each questionnaire question below,

   i   state how it is biased

   ii   rewrite it so that it is not biased.

   a   Do you prefer to holiday in fascinating Sydney or in wet Melbourne?

   b   Do you agree that we should increase the tax on those disgusting cigarettes?

   c   Rate the last movie you saw: OK   Good   Fantastic

   d   Do you prefer exciting rugby league or boring old rugby union?

   e   Do you think the Club is open long enough or should we be able to rage on into the early morning?

   f   Rate your health fund: useless   not much good   reasonable

   g   Are there enough events for young people in this boring town?

   h   Isn't this just the greatest movie you've ever seen?

   i   Are you one of those dumb people who walk to school?

   j   Rate the gym you attend: best ever   good   Ok

2 You have decided to survey your fellow students about the sports offered at your school. Make up 3 examples of biased questions. Then give 3 unbiased versions of these questions.

15.03 Bias in sampling

To be fair and reliable, a sample being surveyed should represent the population accurately. Bias can happen if a sample is not selected properly.

Andrew is considering how to find a fair sample of Club members to complete his survey by:

a   asking 100 people as they come through the door on a Friday night

b   asking 100 people playing Bingo on a Wednesday morning

Shutterstock.com/Blend Images

In what way are these samples biased? How could you ensure that an accurate sample is found?

Solution

a   The people at the club on a Friday night are more likely to be younger people.

All age groups would not be represented fairly.

b   The people at the club on a Wednesday morning are more likely to be retired or not working.

Working members would be under-represented in the survey.

For a more accurate sample, Andrew could find a list of all Uniting Club members and choose every, say, 10th member (other answers are possible).

Exercise 15.03 Bias in sampling

1 For each sample given below, state how:

    i   the sample is biased

    ii   you could ensure that an accurate sample is found.

  a   Asking people in the crowd at an AFL match for their favourite sport

  b   Calling every 6th phone number on a page in the phone book to ask about mobile phone usage

  c   Surveying people about their preferred supermarket in a shopping centre with only one supermarket

  d   Asking people in Brisbane their view on drought support by the Government

  e   Asking teachers what music should be played at the Year 12 formal

  f   Surveying the Maths staff on what books should be bought for the school library

  g   Finding out what people think of a new design for a football club jersey by asking people who walk past the office door

  h   Surveying taxi drivers on who they will vote for in the next election

  i   Emailing a survey to club members to complete

  j   Asking McDonalds customers to name their favourite coffee shop

2 Ms Benedict's Year 11 class is completing a statistical investigation project as part of their assessment. The students need to decide who they will ask to answer their questions. For each investigation, suggest:

    i   what kind of group should be surveyed to get useful results

    ii   what kind of group is likely to give less useful or biased results

  a   New tourist attractions that could be built in Mackay

  b   Children's favourite holiday destinations

  c   Most popular teacher in Nelson Valley State High

  d   Most popular sport in Australia

  e   Changes to the school uniform

  f   Exercise equipment to be installed in outdoor areas

  g   Types of stores to include in a new shopping centre

  h   Entertainment provided in a chain of retirement homes

Chapter problem

You've covered the skills required to solve the chapter problem. Can you solve it now?

TRAVELLING TO SCHOOL

Conduct a survey of how the students at your school commute to school.

1 Before you start, you will need to make some decisions.

- How large a sample will you use?

- How will you select your sample?

- What questions will you ask?

- What are the possible categories of travel that students at your school might use? List them.

- What will you do if there is a category you haven't thought of?

- What will you do if someone uses more than one method of transport?

2 Collect the information.

3 Answer the following questions:

- What is the most common method of travel?

- What is the least common method?

- What effect does the location of your school have on the method of travel?

4 Present your results in a **report**, either written or as PowerPoint slides.

15.04 Errors in surveys

There are different ways in which error can be part of the data collection process. Errors can result from:

- The surveyor not asking the question correctly

- The respondent interpreting the question incorrectly

- The respondent answering the questionnaire incorrectly

- The surveyor recording the answers incorrectly

When collecting data, the aim is to avoid or minimise these errors.

Information that is collected can also be interpreted carelessly to favour a particular point of view.

EXAMPLE 4

Alex stood at the school gate one morning and asked 10 students if they liked the school uniform. Five students said something negative about the uniform. Alex submitted a report to the school principal, claiming that half of the students hate the uniform. Describe 3 things that are wrong with this conclusion.

Solution

1 Alex only asked 10 students and it is not a random selection. The sample is unlikely to be representative of the whole school.

2 The question is vague and it's not clear what students don't like. It may not be the whole uniform that students don't like, but some particular item.

3 Alex exaggerates the answers when he says 'hate'. We don't know how strongly students feel about the uniform.

Exercise 15.04 Errors in surveys

1 Each situation listed below can affect the way a questionnaire is answered. Explain how it could affect the usefulness of the survey results.
   a The questionnaire is very long
   b The presentation of the questionnaire is messy and hard to follow
   c The choice of categories in some questions don't cover all the options
   d The time allowed to answer the questionnaire is limited

2 There are different ways to collect data. For each method listed, describe one possible error to avoid if using that method.
   a Sending a questionnaire by post
   b Asking people to complete a diary each day recording their behaviour
   c Calling up households and asking questions over the phone
   d Having an interviewer ask a person the questions and completing the questionnaire
   e Direct observation, for example, how many vehicles travel through an intersection at a given time of day

3 Jane asks 5 Year 12 students if they like the food served at the school canteen and 3 students say they don't like it. Jane writes an article for the school newspaper with the headline '60% of students hate canteen food'. Explain 3 things that are wrong with this headline.

4 The Hightop Hotel emails its customers to complete a survey about its service. It receives 25 replies, each of which makes some positive comments about the hotel. An advertisement is published saying 'Customers 100% happy with our hotel'. Explain what is wrong with this claim.

5 A newspaper headline reads 'Government program for disadvantaged students a disaster'. In the article, it is revealed that 35 students out of 1000 students receiving government support have dropped out of school. Explain why this headline is misleading.

6 Using the information you collected in the Investigation 'Travelling to school', write a presentation that is deliberately misleading. Perhaps you might like to write about additional student parking, motorbike parking, later school starting times or some other issue, but make sure you distort the true facts!

7 Use the Internet or library to find some newspaper or magazine articles that are based on statistics. Read them carefully to check if they have presented the information properly. Write a report on 3 articles you find, highlighting their use of statistics, either correctly or incorrectly.

INVESTIGATION

STATISTICS IN THE MEDIA

We live in a world of 24-hour news, through TV, news websites, Facebook, Twitter and blogs. To detect bias, we need to consider where the news comes from and what samples were used. Is the information supplied by a journalist, the police, a company executive, a government official or an opinionated blogger? Are the statistics based on a small sample, a large sample, an unrepresentative sample, or a phone/online poll? Each may have a particular bias that may influence how the story is reported.

Find 3 examples of news items or surveys reported in a newspaper or online and investigate the following questions.

1 Where did it come from? What could influence how the story was reported?

2 Who wrote the story?

3 Does it show any bias?

4 Can it be supported by other news providers?

5 What type of sampling was used? Was it representative?

6 How many people were questioned?

7 When was the survey conducted?

Getty Images/Jupiterimages

Questionnaires
word match

Statistical
scramble

KEYWORD ACTIVITY

BIAS IN STATISTICS

1 Explain what is meant by the word *bias* in the statistical context.

2 What are the qualities of a good questionnaire that doesn't have bias?

3 Write a paragraph explaining how samples can be biased and how to avoid this.

4 List 5 errors than can occur in a survey that may lead to unreliable results.

SOLUTION TO THE CHAPTER PROBLEM

Problem

Juanita is researching political views in her community. One Wednesday morning she stands at an intersection on the main street of town. She asks people passing, 'Are you going to vote Liberal or Labor in the next election?'

Will Juanita's data be reliable and unbiased? If not, how can she improve her methods?

Solution

STAGE 1: WHAT IS THE PROBLEM? WHAT DO WE KNOW?

We need to establish whether Juanita's data is reliable and whether there are any improvements she can make to her data collection.

WHAT?

STAGE 2: SOLVE THE PROBLEM

The sample:

SOLVE

On a Wednesday morning she will miss people who are working or shopping. She will mainly be asking older people who have retired, and unemployed people. Her sample is NOT representative of her community.

The question:

There are alternatives to voting Liberal or Labor – other political parties or independent candidates. Her question doesn't cover all the options. Also, she has not specified which election – local council, state or federal governments.

Improvements:

She should choose her sample from the electoral roll of the community.

The question could be asked with a tick box option for answering or it could be an open-ended question. It should specify what level of government is meant. For example, 'Who do you intend to vote for in the next federal election?'

STAGE 3: CHECK THE SOLUTION

CHECK

We have answered all parts of the question. We have used what we know about collecting reliable data.

STAGE 4: PRESENT THE SOLUTION

PRESENT

Juanita's sample is not representative of her community. She should choose a sample from the electoral roll.

Juanita needs to change her question. She must specify what level of politics she is talking about. She can either give a tick box option covering all possible answers or she can make it an open-ended question.

ISBN 9780170412650

15. CHAPTER REVIEW

That's biased

1 Anthony wrote 'How often do you drink?' as part of a questionnaire he was giving to Year 12 students about their alcohol consumption.

  a What problem might people have in answering this question?

  b Rewrite the question, providing some tickbox options.

2 Tranh is surveying students about the quality of food sold at the school canteen.

  a Write an example of a biased question he could use if he wanted to show how great the food is.

  b Rewrite the question you wrote in part **a** so that it is NOT biased.

3 Jenni asks a sample of people in her town, 'Which coffee shop makes the best coffee?'

  a Give an example of a sample that would be biased.

  b State how could you ensure that an accurate sample is found.

4 Kendall interviewed her Year 11 class about changing the times that school starts. 11 of the 20 students surveyed said they wouldn't mind if school started a bit earlier. Kendall presents a report to the student council with the title 'Students support starting school half an hour earlier'. Describe 3 things that are wrong with this claim.

16.

GOING PLACES

Chapter problem

Nina and Michelle are planning a 9 km bushwalk. They plan to walk at a speed of 2.5 km/h over the steep and bushy land. They will leave the car park at 10 a.m. and for safety they will log their walk with the park ranger. At what time should Nina and Michelle tell the park ranger to expect them back at the car park?

WHAT WILL WE DO IN THIS CHAPTER?

- Solve problems involving speed, distance and time
- Interpret distance–time graphs
- Identify the 8 points (directions) on a compass
- Read grid references and information on street maps
- Interpret and use scale on maps
- Find the shortest distance between two places
- Plan trips involving distances, speeds, times and costs

HOW ARE WE EVER GOING TO USE THIS?

- Planning holidays and journeys
- Travelling in unfamiliar locations
- Finding an alternate route to avoid traffic problems

16.01 Distance, speed and time

The question 'How long will it take to get there?' can be answered easily with a little mathematics. The values for the distance covered, the speed and the time taken are related. When we know two of the values, we can calculate the third value.

Speed formula

Distance covered = speed × time

$$D = S \times T$$

The units for speed tell us the units for distance and time.

When the speed is in km/h, the distance is in kilometres and the time is in hours.

Some students find it easier to use the 'distance, speed and time triangle' to solve problems involving speed. Place the letters D for distance, S for speed and T for time in alphabetical order in the triangle.

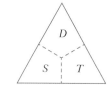

To calculate the **speed**, cover up S, which leaves $\dfrac{D}{T}$.

This means that $S = \dfrac{D}{T}$, or $S = D \div T$.

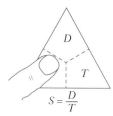

$S = \dfrac{D}{T}$

To calculate the **time**, cover up T, which leaves $\dfrac{D}{S}$.

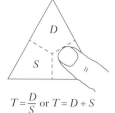

$T = \dfrac{D}{S}$ or $T = D \div S$

To calculate the **distance**, cover up D, which leaves $S \times T$.

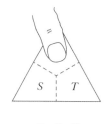

$D = S \times T$

Distance, speed and time

Speed problems

Racing rates

Speed stories

Speed formula practice

Speed, distance and time

EXAMPLE 1

A racing greyhound runs at a speed of 18 m/s.

a   How far will it run in 4 seconds?

b   How long will it take the greyhound to complete a 1200 m race? Answer correct to
the nearest 0.1 s.

Solution

a   Use the triangle. To find the distance,
cover the D.

$D = S \times T$

where $S = 18$ m/s and $T = 4$ s.

$D = S \times T$

$= 18 \times 4$

$= 72$ m

Write your answer.

The greyhound will run 72 m.

b   To find the time, cover the T.

$T = \dfrac{D}{S}$

where $D = 1200$ m and $S = 18$ m/s.

$T = \dfrac{D}{S}$

$D = \dfrac{1200}{18}$

$= 66.666\ldots$

≈ 66.7 s

Write your answer.

The greyhound will complete the race
in 66.7 seconds.

EXAMPLE 2

A kangaroo bounds at a speed of 48 km/h. How far will a kangaroo bound in 20 minutes?

Solution

$D = S \times T$

$S = 48$ km/h

The speed is in km/h, so we need the
time in hours as well. Divide 20 min by
60 to change it to hours.

$T = \dfrac{20}{60} = \dfrac{1}{3}$ h

$D = S \times T$

$= 48 \times \dfrac{1}{3}$

$= 16$ km

Write your answer.

The kangaroo will bound 16 km in 20 min.

EXAMPLE 3

Calvin is driving through a school zone at a speed of 40 km/h.
Convert 40 km/h to a speed in m/s, correct to one decimal place.

Solution

To convert 40 km/h to m/s, change 40 km
to metres and 1 hour to seconds.

$$40 \text{ km} = 40 \times 1000 \text{ m}$$
$$= 40\,000 \text{ m}$$

$$1 \text{ h} = 60 \text{ min} = 60 \times 60 \text{ seconds} = 3600 \text{ s}$$
$$40 \text{ km/h} = \frac{40 \text{ km}}{1 \text{ h}}$$
$$= \frac{40\,000 \text{ m}}{3600 \text{ s}}$$
$$\approx 11.1 \text{ m/s}$$

Write your answer.

40 km/h is equivalent to 11.1 m/s.

A **distance–time graph** is a line graph that describes a journey, by comparing distance with time (on the vertical and horizontal axes respectively). The slope or steepness of the graph indicates speed.

Distance–time
graphs

EXAMPLE 4

This graph shows Monique's cycling trip.

a   At what time did Monique leave home?

b   When was Monique's first stop?
How far from home was she?

c   Find her average speed over the first
2 hours.

d   What time was it when Monique
began her journey home?

e   How far did she travel all together?

f   Find her average speed during the
trip home.

g   For how long did Monique stop
altogether during the trip?

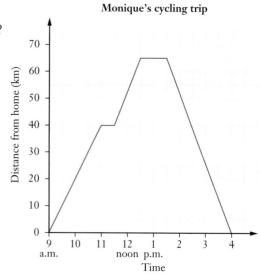

Solution

a Monique left home at the start of the graph, when the distance was 0.

Monique left home at 9 a.m.

b Monique first stopped where the graph is flat. The distance from home does not change, which means Monique has stopped.

Monique first stopped at 11 a.m., 40 km from home.

c Distance = 40 km, time = 2 h.
$$S = \frac{D}{T}$$

$$\text{Average speed} = \frac{40\,\text{km}}{2\,\text{h}}$$
$$= 20 \text{ km/h}$$

d Monique started returning home where the graph points downward.

Monique started returning home at 1:30 p.m.

e Monique travelled 65 km, then returned home.

$$\text{Total distance} = 2 \times 65 \text{ km}$$
$$= 130 \text{ km}$$

f Distance = 65 km, time = $2\frac{1}{2}$ h
$$S = \frac{D}{T}$$

$$\text{Average speed} = \frac{65\,\text{km}}{2\frac{1}{2}\,\text{h}}$$
$$= 26 \text{ km/h}$$

g Look at the places where the graph is flat.

First stop: $\frac{1}{2}$ hour.

Second stop: 1 hour

Total stopping time $= \frac{1}{2} + 1 = 1\frac{1}{2}$ hours

On a distance-time graph:

* a horizontal (flat) section on the graph indicates a stop

* the steeper the line, the greater the speed (more distance covered in less time)

* a section going down, towards the right, indicates a change in direction or that the traveller is returning towards the start.

ISBN 9780170412650

Exercise 16.01 Distance, speed and time

1 a Use the formula $T = \dfrac{D}{S}$ to determine the value of T when $D = 80$ and $S = 16$.

   b In the formula $D = S \times T$, what is the value of D when $S = 80$ and $T = 3$?

2 a Corrina is driving at a speed of 60 km/h. How far will she drive in 3 hours?

   b How long will it take her to drive 240 km?

3 Anton has an appointment 160 km away. He must be there in 2 hours. At what speed must he travel to arrive in time?

4 Go-karts can race at a speed of 110 km/h. At this speed, how many kilometres can a go-kart travel in a $2\frac{1}{2}$-hour race?

5 Wasim is driving through heavy traffic at a speed of 32 km/h. How far will he travel in 15 minutes?

6 An ambulance is racing to the scene of a serious freeway accident, at a speed of 100 km/h. The accident is 15 km from the ambulance station.

   a How long will the ambulance take to reach the accident? Express your answer as a decimal of an hour.

   b Multiply your answer to part **a** by 60 to change the time to minutes.

7 A tactical response team is travelling at 80 km/h to reach a hostage situation 10 km away. How long will it take the team to arrive at the scene? Express your answer in minutes.

8 A whitewater rafting team completed 3 sets of rapids and 1.6 km of calm water in 30 minutes. The lengths of the sets of rapids were 150 m, 80 m and 170 m.

   a Calculate the distance that the rafting team covered in 30 minutes. Express your answer in kilometres.

   b Explain why you can't use $T = 30$ in the equation $S = \dfrac{D}{T}$ to calculate the speed of the raft in km/h.

   c Calculate the raft's average speed in km/h.

9 Cheetahs are the fastest animals on land, and can run at a speed of 31 m/s in short bursts.

   a How far can a cheetah run in 9 seconds?

   b How long does it take a cheetah to run 140 m? Answer in seconds, correct to 1 decimal place.

10 A peregrine falcon's top speed is 90 m/s.

   a How far can the falcon fly in one minute?

   b How long will it take the falcon to fly 1 km?

11 The table shows the distances between several cities in kilometres.

	Albury	Brisbane	Canberra	Goulburn	Sydney	Tamworth
Albury	–	1610	190	380	600	1040
Brisbane	1610	–	1300	1225	1020	575
Canberra	190	1300	–	95	300	750
Goulburn	380	1225	95	–	205	660
Sydney	600	1020	300	205	–	460
Tamworth	1040	575	750	660	460	–

a   How far is it from Canberra to Tamworth?

b   How long will it take to drive from Canberra to Tamworth at an average speed of 75 km/h?

c   Glen took 5 hours to drive from Goulburn to Albury. What was his average speed?

d   Max and Sanjay left Brisbane at 6 a.m. on Monday to drive to Albury. They shared the driving and completed the trip at an average speed of 70 km/h. At what time did they arrive in Albury?

12 Convert each speed to m/s.

a   60 km/h   **b**   90 km/h

Example 3

13 The minimum speed a rocket needs to escape the Earth's gravity is 11.2 km/s. What is this speed in km/h?

14 a   Use a stopwatch to time how long it takes you to walk across your classroom.

b   An international space station travels at a speed of approximately 28 000 km/h. How far does the space station travel in the time it takes you to walk across your classroom?

PS

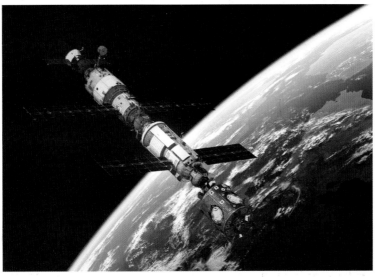

Example 4

15 Kyle and Ashley went to the cinema together. The distance–time graph shows the distance they were from Kyle's house after the movie finished.

a How far is the cinema from Kyle's house?

b After the movie finished, Kyle and Ashley rode their bikes to Ashley's house. How far is Ashley's house from the cinema?

c How long did it take them to ride from the cinema to Ashley's house?

d What was their average speed riding to Ashley's house?

e How long did Kyle stay at Ashley's house?

f Calculate Kyle's average speed when he was riding home.

g Part of the trip home is downhill and part of it is uphill. Did he ride the uphill section during the first or second hour of his ride home? Justify your answer.

16 This graph shows a cyclist's day trip.

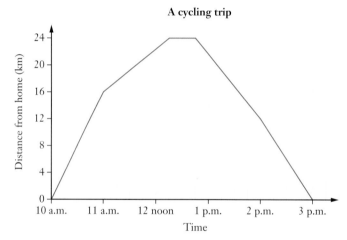

a At what time did the speed of the cyclist

i increase? **ii** decrease?

b When did the cyclist start to return home?

c How far did the cyclist travel altogether on this day?

d How long did the cyclist spend 'on the road'?

e Find the cyclist's average speed for:

i the first hour **ii** 11:00 a.m. to 12:15 p.m.

iii 12:45 p.m. to 2:00 p.m. **iv** the entire day

17 Brian travels from Bligh to Macquarie, while Sam travels from Macquarie to Bligh.

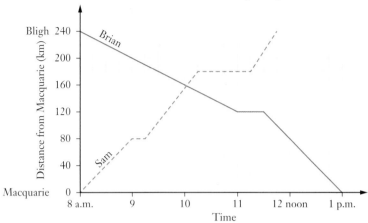

Brian and Sam's journey

a How far is it between the 2 towns?

b Who is travelling faster? How can you tell?

c How far is Brian from Macquarie at 11 a.m.?

d How far is Sam from Macquarie at 11 a.m.?

e At what time do Brian and Sam pass each other? How far are they from Macquarie when they pass?

f Is Sam travelling faster before 9 a.m. or after 9:15 a.m.? How does the graph show this?

g Calculate Brian's average speed before he stops.

h For how long did Sam stop altogether on the trip? Select the correct answer **A**, **B**, **C** or **D**.

    A 60 minutes   **B** 75 minutes   **C** 5 minutes   **D** 90 minutes

18 Use the clues to determine which person matches with each graph. Copy and complete the table on the next page to record your answers.

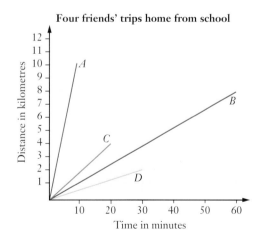

Four friends' trips home from school

Clues

- Shelby walks home at a speed of 4 km/h.

- Wayne rides his bike at an average speed of 12 km/h.

- Peta travels on the slow bus. With all the stops, the bus averages only 8 km/h.

- Luke drives home at an average speed of 60 km/h.

Graph	Person	Distance from school
A		
B		
C		
D		

19 Construct a graph, similar to the graph in question **18**, which shows the trips home from school made by you and 3 other people in your class.

INVESTIGATION

PROBLEM SOLVING: THE GREAT TRAIL BIKE RACE

Renaldo and Megan were racing against each other in a trail bike competition.

Renaldo completed the first half of the course at an average speed of 5.5 m/s and the second half at an average speed of 4.5 m/s.

Megan rode at an average speed of 5.0 m/s for the whole race.

1 Which rider completed the race in the faster time? Justify your answer.

2 A common misconception is that Renaldo's average speed is (5.5 + 4.5) ÷ 2 = 5.0 m/s, which is the same speed as Megan, so they should finish the race at the same time. Write a paragraph explaining why this thinking is wrong.

> Trying a few different values for the length of the race, such as 4000 m or 1000 m, could help you solve the problem.

Imagefolk/Marc Oeder/Westend61

SPEED UNITS

Use the Internet to investigate a variety of speeds shown in the table below.

Decide on the appropriate units to express the speed: m/s, km/h, km/s, m/h or more than one. Develop a rule for answering the question: 'What units should I use?'

Copy and complete this table or print a copy from NelsonNet.

Speed units

Activity	Typical or record speed
Slow speeds	
Moving snail	
Adult walking	
Athlete running	
Medium speeds	
Cruise ship	
Car	
Downhill cross-country skiing	
Olympic speed-skating	
Fast speeds	
Queensland tilt train	
Japanese high-speed train	
International airline jet	
Very fast speeds	
Bullet fired from an assault weapon	
Fighter jet	
International Space Station	
Sound	
Light	

Group discussion questions

1 In what situations do we use the following speed units: m/h, m/s, km/s, km/h?

2 Why might we decide to measure the same speed using different speed units?

Chapter problem

You've covered the skills required to solve the chapter problem. Can you solve it now?

16.02 Compass directions

The 4 major compass directions are north (N), east (E), south (S) and west (W). The directions halfway between these are northeast (NE), southeast (SE), southwest (SW) and northwest (NW). We use these 8 compass directions to describe wind direction, position of the Sun and shadows, and the path of a plane or ship.

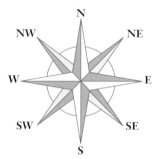

EXAMPLE 5

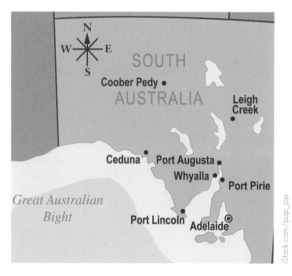

a   What compass direction is it from Coober Pedy to Port Augusta?

b   In what direction is Coober Pedy from Port Augusta?

Solution

a Position the centre of a compass rose on the location the direction is *from*. In this case, the bearing is from Cooper Pedy.

The direction to Port Augusta is in the middle of east and south. The direction from Cooper Pedy to Port Augusta is southeast.

b Now the bearing is from Port Augusta. Position the compass rose on Port Augusta.

The direction to Cooper Pedy is between north and west. Cooper Pedy is northwest of Port Augusta.

ISBN 9780170412650

Exercise 16.02 Compass directions

Example 5

1 What is the compass direction:

   a from Hobart to Roseberry?

   b from Launceston to Bothwell?

   c to Roseberry from Queenstown?

   d to Launceston from George Town?

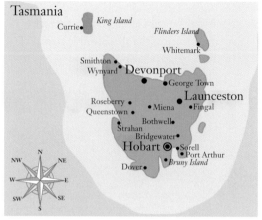

2 In 1986, following an explosion and fire at the Chernobyl nuclear power station, winds carried the radiation cloud over a vast area.

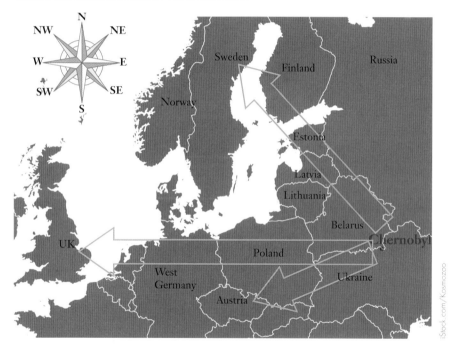

   a What compass direction, NW, NE, SE or SW is it from Chernobyl to Austria?

   b Easterly winds are winds that come from an easterly direction. What countries experienced radiation clouds as a result of easterly winds?

3 When Muslims pray, they face the central shrine in Mecca's Great Mosque, Islam's holiest place. What is the approximate compass direction (N, S, E, W, NE, SE, SW or NW) from each of the following places to Mecca?

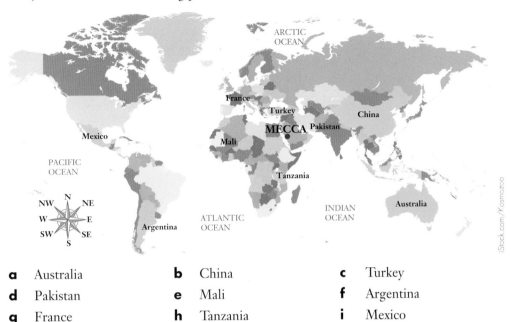

a	Australia	**b**	China	**c**	Turkey
d	Pakistan	**e**	Mali	**f**	Argentina
g	France	**h**	Tanzania	**i**	Mexico

4 Late in the afternoon, the Sun is in the west. Which one of the diagrams shows the correct position of the tree's shadow?

PS

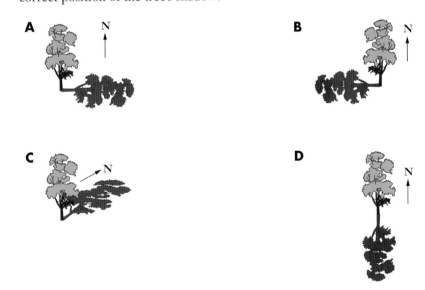

5 A westerly wind brought smoke from a
bushfire to Parkes.

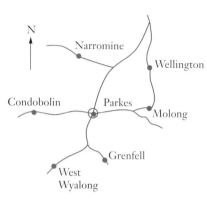

> A westerly wind comes from the west.

a Near which town on the map could the
bushfire be located?

b If the wind keeps blowing in the same
direction, which other town will receive
smoke from the fire?

c The wind changed direction and smoke went
in the direction of SE.
Which town experienced smoke?

6 Winds from over the ocean carry moisture and are likely to bring rain, but winds from
over a land mass are likely to be dry. The temperature of the land the winds blow over
determines whether the winds will be hot or cold. In winter, winds from over the snowfields
will be cold, but summer winds from Central Australia are likely to be dry and hot.

Are the following winds likely to be hot, cold, bring rain or fine weather?

> Remember! A wind's direction is the direction it has come *from*.

a westerly winds in Perth

b westerly winds in Bathurst in summer

c northeasterly wind in Goulburn

d southerly winds in Canberra in winter

e southerly winds in Darwin in summer

f southerly winds in Adelaide

g northwesterly winds in Broome

h southeasterly winds in Broome in summer

16.03 Street maps

Street maps have a grid coordinate system to help us locate places. Usually we read the **coordinates** by starting with the horizontal (across) letters, then the vertical (up and down) numbers.

This is a street map of Manly in Sydney.

a On this map, what is the name of the park at F3?

b What are the position coordinates of Manly Oval?

c What direction is it from Manly Oval to Manly Wharf?

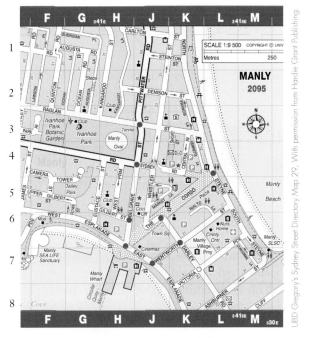

Solution

a Find the position where column F meets row 3.

Ivanhoe Park is located at F3.

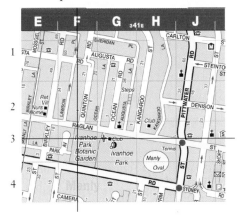

16. Going places

b   Manly Oval is in column H and rows 3 or 4.

Manly Oval is at H3 or H4.

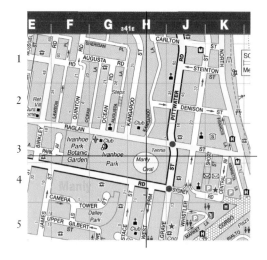

c

This symbol 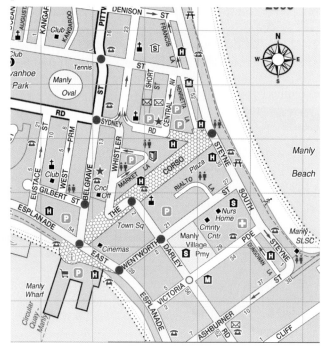 shows the compass directions on the map. On this map, North (N) is straight up.

An imaginary line from Manly Oval to Manly Wharf points south.

Manly Wharf is south of Manly Oval.

Distances on a street map

Street maps have a scale, which is shown on the map as a diagram or ratio. The scale on the Manly map is 1 cm represents 95 m, also shown as the ratio 1 : 9500 or as a diagram.

SCALE 1:9 500

| Metres | 250 | 500 |

EXAMPLE 7

On this street map of the Melbourne CBD, each grid square represents a distance of 62.5 m. Use this scale to calculate the length of the Royal Melbourne Institute of Technology campus along Swanston St.

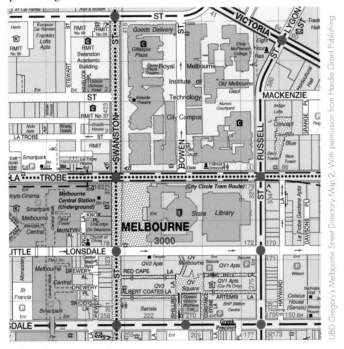

UBD Gregory's Melbourne Street Directory Map 2. With permission from Hardie Grant Publishing.

Solution

The length of the campus along Swanston St covers 4 grid squares. Multiply by 62.5 m.

Write your answer.

Boundary length = 62.5 × 4

= 250 m

The Swanston St boundary is 250 m long.

If we need to calculate a distance that isn't along the side of a grid square we can place a chain or piece of string along the distance, then compare the distance to the sides of the grid. In this picture, the length of the chain is roughly 3 grid squares.

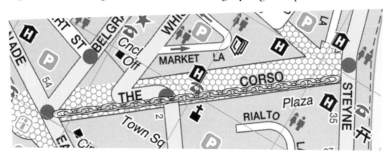

Exercise 16.03 Street maps

Use map 17 next page to answer questions **1** to **8**.

Example
6

1 a What is the name of the school at F9?

b What features indicated by small black squares are located in B5?

c Which shopping centre is located at H12 and H13?

d There is an ambulance station at J18 on the side of Kingston Drive. What symbol represents an ambulance station?

2 State the position coordinates of each place.

a Westfield Helensvale

b Warner Bros Movie World

c Aveo Tranquility Gardens

3 Name the compass direction from:

a Westfield Helensvale to Jubilee Primary School (A14)

b Jubilee Primary School to Warner Bros Movie World

c Helensvale Primary School to Arcare Helensvale (F12, F13)

d Arcare Helensvale to Helensvale Plaza

Example
7

4 On the map, 1 grid square represents a length of 250 metres. Use the grid squares to find:

a the distance between the traffic lights on Discovery Drive, near the community centre (J10) to the traffic lights at the intersection of Discovery Drive and Lindfield Rd (H11).

b the direct distance from the roundabout at Helensvale High School (E1) to Helensvale railway station (K13).

c the length of the Pacific Motorway from the on-ramp in A5 to the off-ramp in E13.

d the length of the bike track from the Philip Grey roundabout at (H15) to the traffic lights at Warner Bros Movie World (A6).

MAP 17
1 KILOMETRE EQUALS 4 GRID SQUARES

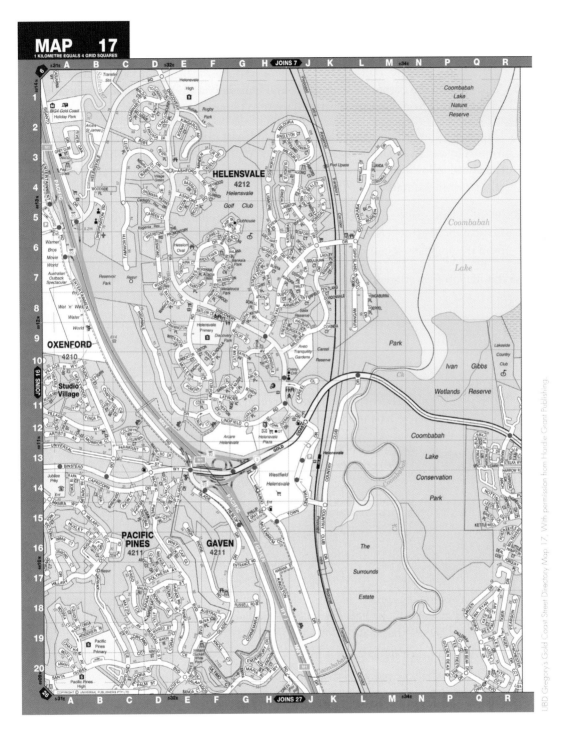

5 After Kelly got off the train at Helensvale station, she walked home. She walked north along Town Centre Drive. She crossed over the Gold Coast Highway and turned into the second street on the left, then the first on the right. Where does Kelly live?

6 Anna lives on the corner of Ashford Rd and Quirindi Ct (F4). Write instructions for her friend to walk from Hession Oval (E6) to Anna's house.

7 Dinesh is trying to drive west along the Gold Coast Highway to join the Pacific Motorway and head north towards Brisbane. At present he is at K11. There is a traffic jam ahead following a serious road accident. Describe an alternate route Dinesh could take to avoid the traffic.

8 An ambulance has been called to attend the accident on the Gold Coast Highway in front of the Westfield Shopping Centre. Describe a route for the ambulance to follow from the ambulance station in J18 to the accident in H13.

Use this map to answer questions **9** and **10**.

9 Katarina lives on the corner of Port Hacking Rd and Percival Rd (position of the red dot). She walks southeast along Port Hacking Rd and Parthenia St to Zhi's house on the corner of Wisteria St and Parthenia St.

   a What compass direction is it from Katarina's house to Zhi's house?

   b Place a chain or piece of string on the map along the Katarina's route and use the scale on the map to calculate the approximate distance she walks.

10 The next day, Katarina walks northwest from her home to the corner of Mirral Rd and Blamey Ave.

   a Describe the shortest route she could take.

   b Approximately how long is this route?

   c At an average speed of 4 km/h, approximately how many minutes will the walk take her?

DIRECTIONS TO YOUR HOME

Imagine your friends are going to visit your home but they have never been there before. They will be arriving at the closest train station, bus stop, ferry terminal or airport and then require directions to your place.

What you need to do

- Use either a street map or Google Maps to write instructions for your friends to walk, ride a bike or catch a taxi to your place.

- Include distance and direction indications such as left and right as well as compass directions to make following your instructions easy.

16.04 Scales on maps

In Chapter 6, we learned about scales and scale drawings. We can use a map's scale to calculate lengths on the map.

Map scales

The scale on a bushwalking map is 1 : 5000. On the map, a walking trail is 4.5 cm long. How long is the trail in real life?

Solution

1 : 5000 means that 1 unit on the map means 5000 units in real life. Multiply 4.5 cm on the map by 5000 to find the real length.	Real length $= 4.5 \times 5000$ $= 22\,500$ cm
Convert to metres.	$= 22\,500 \div 100$ m $= 225$ m
Write your answer.	The trail is 225 m long.

Exercise 16.04 Scales on maps

1 A map's scale is 1 : 2000. Calculate the real lengths of each feature.

  a   Lake 1.5 cm wide

  b   Walking track 5.8 cm long

  c   Creek 9.5 cm long

Example 8

---

2 The map shows the walking trails that start at the lodge.

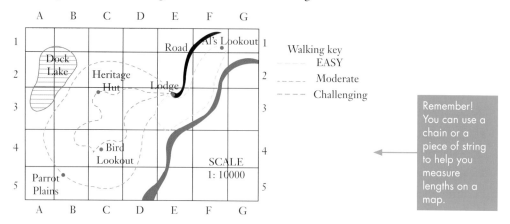

a What is the scale?

b What length in real life is represented by 10 cm on the map?

c How long is the creek?

d Calculate the length of the moderate walking track.

e Jude started walking along the moderate walking track at 11 a.m. She is walking at a speed of 3 km/h. When will Jude finish walking the track?

3 Zack lives in Corona Lane. The star on the map shows the position of his house.

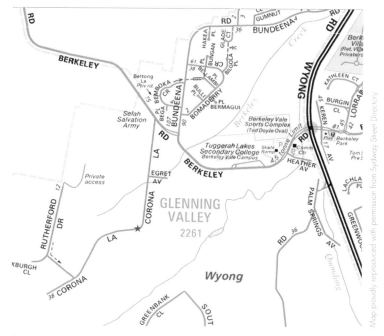

a Use the scale 1 : 10 000 to calculate the distance Zack walks along roads to Tuggerah Lakes Secondary College each day.

b Zack walks to school at a speed of 1 m/s. How long does it take him to walk to school?

4 This map shows the road around East MacDonnell National Parks in Central Australia. The scale is 1 : 1 000 000.

   a Use the scale to calculate the length of the road (the circuit outlined in dark blue).

   b The suggested time to drive one circuit of this road is $5\frac{1}{2}$ hours. At what average speed do drivers need to travel to complete the circuit in $5\frac{1}{2}$ hours? Express your answer correct to the nearest km/h.

   c What do you think the surface of the road is like? Give a reason for your answer.

5 **Cadastral maps** show the boundaries of properties. This part of a cadastral map shows the dimensions of a block of land and the buildings on the land.

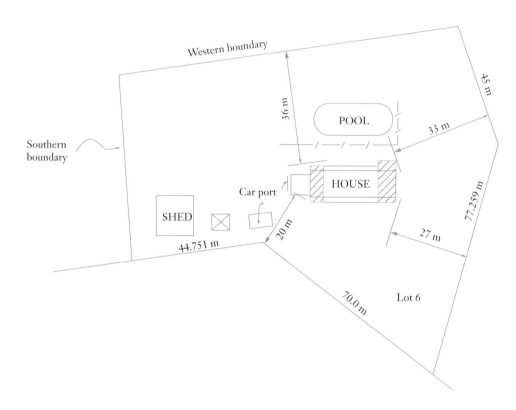

a Approximately how close to the house is the nearest boundary?

b The scale is 1 : 1000. Calculate the length of the southern and western boundaries.

c The owners are going to replace the fence on the southern and western boundaries. The fencing costs $28/m for materials and labour. Calculate the cost of replacing the fencing.

6 The town of Gympie in South-East Queensland is prone to flooding. The map shows the areas affected when the height of the Mary River is 18 m and 23.5 m at the Kidd Bridge (not shown on the map but south of the map). The scale is 1 : 2300.

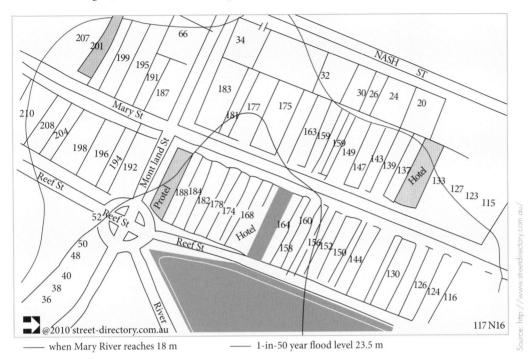

——— when Mary River reaches 18 m   ——— 1-in-50 year flood level 23.5 m

Source: http://www.streetdirectory.com.au/

a Which properties in Mary Street flood when the Mary River reaches 18 m?

b Calculate the length of Reef St that is under water when the Mary River is at 18 m.

c Calculate the length of Mary Street that is under water during a 1-in-50 year flood.

d Estimate the height of the Mary River when the shop at 187 Mary St floods.

e Rhys is considering buying the hotel at 135 Mary Street. What recommendation could you give him?

7 Jose, Isaac and Chase were hiking along the coastal track in the National Park. They left their bikes at Otford lookout and set out for Garie Beach.

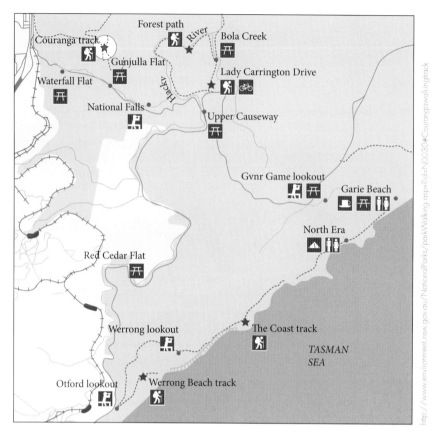

a The map scale is 1 : 87 600. How long is the walk from the Otford lookout to Garie Beach?

b The boys hike at an average speed of 4 km/h. How long should the hike take them?

c When they had been hiking for 2 hours, Chase lost his footing, fell and broke his leg. Isaac phoned emergency services for help. Approximately where on the track should he tell the rescue helicopter they are?

INVESTIGATION

A CADASTRAL MAP OF YOUR SCHOOL

A cadastral map shows information about the boundaries of individual and surrounding blocks of land. In this investigation, you are going to make a cadastral map of your school and surrounding properties. Make the map suitable for display in the school's front office foyer.

On the map you should include a compass rose, a scale, the lengths of all boundaries and the position of buildings.

Start your investigation by downloading a map and a picture of the terrain from Google Earth.

16.05 Distances between places

We can use roughly-drawn diagrams called **mud maps** to represent roads. Here is an example of a map of some roads linking towns A, B, C and D, with all the measurements written in kilometres.

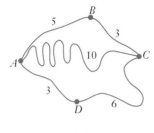

Map

EXAMPLE 9

a In the mud map above, list 3 different ways to travel from towns A to C and the length of each route.

b Which route is the shortest?

Solution

a We could travel directly from A to C, which is 10 km. We could also go via B or D. We can calculate the distances via B or D by adding the distances on the roads.

There are 3 routes:

$AC = 10$ km

$ABC = 5 + 3 = 8$ km

$ADC = 3 + 6 = 9$ km

b Compare the three distances and determine the shortest route.

$ABC = 8$ km.

The shortest routes from A to C is via B.

Exercise 16.05 Distances between places

In this exercise, all measurements are in kilometres.

1 a List 3 different routes from A to D.

b Calculate the length of each route from A to D.

c Which route from A to D is the shortest?

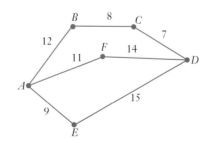

Example 9

2 a List 3 possible ways to travel from *A* to *B* and the distance involved with each route.

b How long is the route from *A* to *B* via *C* and *D*?

c What is the shortest distance from *A* to *B*?

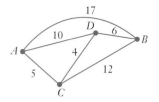

3 a List 4 ways you could travel from *A* to *C*.

b Which one of the 4 ways from *A* to *C* is the longest?

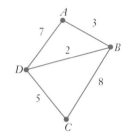

4 Jack is visiting a wine region of the Darling Downs. He sketched a mud map of the places he wants to visit. The numbers are the distances in kilometres.

a Jack is at the chocolate shop. What is the shortest way to the restaurant?

b What is the shortest distance from the restaurant to the winery?

c Calculate the shortest distance from the winery to the cheese factory.

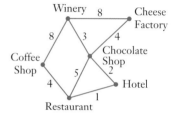

5 The local fire station has a mud map of the area to help get fire trucks to locations efficiently.

There are fires at The Ridge and One Hill Lookout. Describe the shortest routes for the fire trucks to get to the fires.

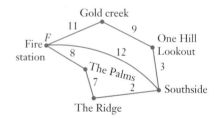

Imagefolk/FLPA/John Eveson

6 Use the map of Manly on page 381 to find the shortest way by road from the corner of Victoria Parade and East Esplanade to Manly Oval in Raglan Street. Check your answer using Google Maps.

INVESTIGATION

SHORTEST DISTANCE FROM HOME

You will need a street map of your local area to complete this investigation.

Locate your home and a useful location on the map, for example, hospital, bank, library, police station or pool.

Determine the shortest way to get from your home to the location you selected travelling on roads. Check your route using Google Maps.

16.06 Regional maps

Josie is travelling from Tamworth to Walcha. How far is it by road?

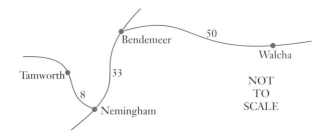

It's 8 km from Tamworth to Nemingham, then 33 km to Bendemeer and 50 km to Walcha.

Total distance = 8 + 33 + 50 km

= 91 km

Write your answer.

It is 91 km from Tamworth to Walcha.

Solution

Exercise 16.06 Regional maps

Use this map from Tasmania to answer questions **1** to **4**.

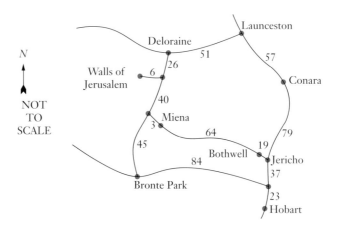

1 Terry drove from Launceston to Bothwell via Jericho. How far did he drive?

2 How far is it from Launceston, via Deloraine, to Miena?

3 Luana drove from her hotel in Miena to see the Walls of Jerusalem and she drove back to her hotel for lunch. What distance did she travel on the round trip?

4 When Luana left Miena to travel to Hobart, the odometer on her car showed $\boxed{3}\boxed{5}\boxed{2}\boxed{1}$ She drove south-east and then south.

   a What was the reading on her odometer when she reached Hobart?

   b The trip took 2 hours and 15 minutes. Calculate her average speed from Miena to Hobart, correct to the nearest km/h.

5 Mick left Emerald to go sapphire mining in the nearby gemfields.

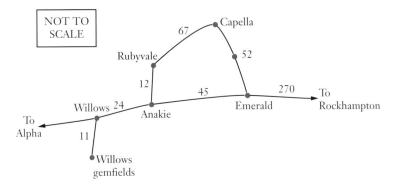

   a How far is it from Emerald to the Willows gemfields?

   b How long will it take Mick to drive to the Willows gemfields at an average speed of 75 km/h?

   c Sapphire mining is very hard work in the sun. Mike didn't take sufficient water with him and the closest shop that sells water is in Anakie. At 1:10 p.m., he left the gemfields to get water. If he averages 70 km/h, what time should he arrive at the shop?

   d Mike found a big star sapphire in the gemfields and was very excited! He drove to Rubyvale in 20 minutes to have it valued. He was booked by a speed camera for travelling 120 km/h in a 100 km/h zone.

   i Calculate Mike's average speed from the gemfields to Rubyvale.

   ii So how can the speed detected by the speed camera be explained?

6 Linda drove from Darwin to Katherine in the Northern Territory and then on to see the spectacular red sandstone cliffs in Keep River National Park.

When she left Darwin, the car's odometer reading was ⌗ 7 ⌗ 3 ⌗ 4 ⌗ 7 ⌗ 5 ⌗, and when she reached the Keep River National Park, it was ⌗ 7 ⌗ 4 ⌗ 2 ⌗ 6 ⌗ 1 ⌗.

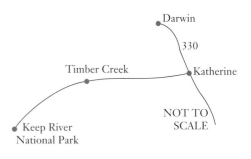

a How far is it from *Katherine* to the Keep River National Park?

b Linda drove at an average speed of 60 km/h. How long did the drive take her?

7 Claire and Jimmy drove southwest from Sydney for 123 km and stopped for lunch.

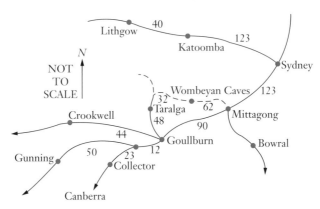

a Where did they stop for lunch?

b Wombeyan Caves was their next stop. They thought the dirt road directly to the caves looked too rough, so they went via Goulburn. How far did they travel?

c Their average speed after lunch was 90 km/h. How long did it take them?

d Their friends Jorja and Gilbert met them at the caves. They drove the 'short', dirt road route from Mittagong to the caves. It took them 1 hour and 40 minutes. What was their average speed?

8 At Wombeyan Caves, Claire and Jimmy camped in their tent while Jorja and Gilbert stayed in an on-site caravan.

Accommodation charges

- **Cabin units** $85 per night
- **On-site caravan** $72 per night
- **Guest cottage** $124 per night
- **Caravan park site** $25 per night per site
- **Camp site** $19 per site per night

Calculate the total accommodation costs for the 4 friends to stay for 2 nights.

9 a Claire and Jimmy went on an afternoon tour of the Junction Cave. At what time did the tour start?

Cave tours times

Fig Tree cave (Self guided)	9 am to 5 pm
Wollondilly	1:45 pm
Junction	11 am and 2:30 pm
Mulwaree	11:30 am
Kooringa	3:15 pm

Cave inspection fees

	Adult	Child	Family
Self guided (Fig Tree cave)	$15	$7	$35
Guided tour	$18	$9	$40
Two-cave package	$22	$12	$50

b How much did the tour cost?

c Jorja and Gilbert toured both the Wollondilly and Fig Tree Caves. How much did they save altogether by buying the two-cave package instead of separate tickets?

10 The next day, the 4 friends had breakfast in Goulburn. Then they left Goulburn at 10:15 a.m. and drove at an average speed of 85 km/h to Gunning. What time did they arrive in Gunning?

16.07 Family road trip

The Winslow family of 2 adults and 3 children are travelling from Brisbane to South Australia for a holiday. They plan to stay in motels, caravan parks and the occasional luxury hotel.

Exercise 16.07 Family road trip

1 Buses travel from Brisbane to Adelaide every day. Adelaide's local time is 30 minutes behind Brisbane and the times in the bus timetable are local times.

Location	Arrival time	Departure time
Brisbane (Roma St)		8:00 am
Goondiwindi	12:15 pm	1:00 pm
Dubbo	6:40 pm	7:00 pm
Wilcannia	1:00 am	1:15 am
Cockburn	3:47 am	
	Time change	
Burra	7:33 am	8:13 am
Adelaide		

a The family plan to arrive at Roma St bus terminal 30 minutes before the bus is due to leave. What time did they plan to be at Roma St?

b The bus stopped at Goondiwindi for the passengers to have lunch. How long was the lunch stop?

c How long does the bus take to travel from Dubbo to Wilcannia?

d Calculate the average speed of the bus between Dubbo and Wilcannia, a trip of 557 km. Answer correct to the nearest km/h.

e Cockburn is on the NSW/SA border and the bus doesn't stop in Cockburn. What is the South Australian local time as the bus travels through Cockburn?

f The bus stops in Burra for the passengers to have breakfast. The family ordered breakfast and it arrived at 7:55. How long do they have to finish it?

g The trip from Burra to Adelaide takes 2 hours and 7 minutes. At what time does the bus arrive in Adelaide?

h How long does it take to travel from Brisbane to Adelaide on the bus?

i The adult bus fare is $210 each way and children pay 70% of the adult fare. How much did it cost the Winslow family of 2 adults and 3 children to travel from Brisbane to Adelaide on the bus?

j The cheapest budget airline tickets from Brisbane to Adelaide are $170 each and the flight takes approximately 8 hours, including a stopover. How much more does it cost the family to travel by bus to Adelaide compared to flying? Suggest some reasons why the family might have decided to go by bus.

2 From Adelaide, the family visited Kangaroo Island in South Australia. They stayed here for 2 days and enjoyed watching its friendly wildlife.

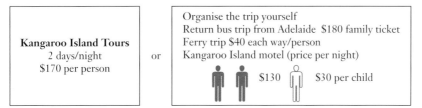

How much did the family save by using Kangaroo Island Tours instead of organising the trip themselves?

3 The Winslows returned to Adelaide and hired a 4-wheel-drive vehicle to travel north to Peterborough, and then Quorn.

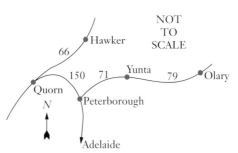

a What compass direction is it from Peterborough to Quorn?

b They left Peterborough at 11 a.m. and arrived at Quorn at 1 p.m. Calculate their average speed.

4 The Winslows stayed at the Quorn Caravan Park.

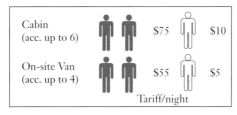

a How much did it cost the family to stay in the cabin for 2 nights?

b How much more would it have cost them to use the on-site vans for 2 nights?

c Why would it have cost more for the Winslow family to stay in the on-site vans rather than the cabin?

d Will it always be more expensive to stay in vans than in cabins? Explain your answer.

5 The region around Quorn is shown on this map.

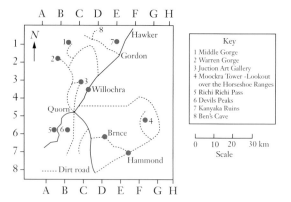

a What scenic attraction is at F6?

b What is the name of the town at D7?

c The children wanted to see some kangaroos. The park manager gave them these directions to a place frequented by kangaroos:

'Travel north-east from Quorn for 24 km and turn left onto a dirt track. Travel another 24 km, then take the track to the right.'

Where did they go to see the kangaroos?

d Mrs Winslow wanted to see the Horsehoe Ranges. She asked for directions to get there from Quorn. What could the directions have been?

6 The next part of the road trip involved travel from Hawker to Arkaroola.

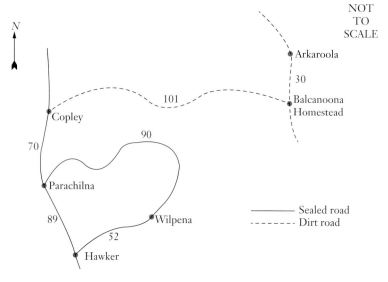

a How far is it from Hawker to Arkaroola via the shortest route?

b The recommended average speed from Hawker to Copley is 80 km/h and from Copley to Arkaroola is 40 km/h. The Winslows followed the recommended average speeds. How long did it take them to drive from Hawker to Arkaroola?

7 At Arkaroola, the Winslows had a choice of 2 luxury lodges.

Greenwood Lodge	Room rates per night
Basic charge	$199 (per double or single)
Additional adult	$36 (each)
Children (sharing) with adults	$30 (each)
Child aged 2 to 12	1 double bed and
Family room size	1 single bed

Mawson Lodge	Room rates per night
Basic Charge	$219 (Per double or single)
Additional adult	$36 (each)
Children (Sharing with adults)	$30 (each)
Child aged 2 to 12	1 double bed and
Family room size	3 single beds

The Winslow children are aged 13, 11 and 8.

a Explain why Mawson Lodge is the more suitable accommodation for the family.

b Calculate the cost of one night's accommodation for the family in Mawson Lodge.

8 The family decided to fly from Arkaroola to Adelaide. The flight takes 2 hours and their plane back to Brisbane leaves Adelaide at 1700. Flights leave Arkaroola at 0600, 0715, 0900, 1345, 1445 and 1600.

a What flight do you recommend they catch from Arkaroola to Adelaide to be on time for their flight home to Brisbane? Give a reason for your answer.

b The flight from Adelaide to Brisbane takes 2 hours and 40 minutes. Calculate the local time when the flight arrives in Brisbane.

INVESTIGATION

WHERE IS ARKAROOLA?

Alamy Stock Photo/David Foster

What is there to see in Arkaroola? Use the Internet to find out some interesting information about Arkaroola and prepare a travel poster to advertise the best features of this location.

Copy this crossword, then use the clues to complete it.

Across

5 The letter-number combination that gives the position of a street or object on a street map

8 What D stands for in the formula $D = S \times T$

9 A type of direction, for example, north, south, east and west

Down

1 What 's' stands for in the speed units m/s.

2 Speed is often measured in _____ per hour.

3 On a map, the direction pointing down the page

4 A ratio that allows us to calculate real distances from a map

6 The opposite direction to southeast

7 Distance ÷ time

10 A scale diagram of a city or region

SOLUTION TO THE CHAPTER PROBLEM

Problem

Nina and Michelle are planning a 9 km bushwalk. They plan to walk at a speed of 2.5 km/h over the steep and bushy land. They will leave the car park at 10 a.m. and for safety they will log their walk with the park ranger. At what time should Nina and Michelle tell the park ranger to expect them back at the car park?

Solution

WHAT?

STAGE 1: WHAT IS THE PROBLEM? WHAT DO WE KNOW?

To work out when the bushwalk will finish.

Leaving 10 a.m.

Walking at 2.5 km/h

9 km walk

SOLVE

STAGE 2: SOLVE THE PROBLEM

Use a formula to calculate time taken.

$$T = \frac{D}{S}$$
$$= \frac{9}{2.5}$$
$$= 3.6 \text{ hours}$$
$$= 3 \text{ hours } 36 \text{ minutes}$$

Nina and Michelle will start the walk at 10 a.m. and finish 3 h 36 min later. They will finish at 1:36 p.m.

CHECK

STAGE 3: CHECK THE SOLUTION

Walking 2.5 km/h means covering 5 km in 2 hours and 10 km in 4 hours, so a 9 km walk in 3.6 hours sounds reasonable. The solution sounds correct.

PRESENT

STAGE 4: PRESENT THE SOLUTION

Nina and Michelle should tell the park ranger to expect them to return before 2 p.m.

Going places

Exercise
16.01

1 How long will it take Bianca to ride 24 km at an average speed of 8 km/h?

Exercise
16.01

2 Kane is driving at an average speed of 60 km/h. How far will he drive in 15 minutes?

Exercise
16.01

3 Ava travelled 900 m in 45 seconds. Calculate her average speed in m/s.

Exercise
16.01

4 Convert a speed of 90 km/h to a speed in m/s.

Exercise
16.01

5 The graph shows the distance that Kelly was from home. What was Kelly's average speed?

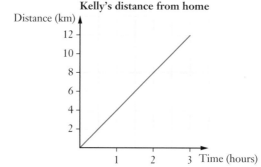

Kelly's distance from home

Exercise
16.01

6 Guy is riding his bike. For the first 3 hours, he rode at 12 km/h and for the last 2 hours he rode at 8 km/h.

   a Explain why the calculation $(12 + 8) \div 2$ doesn't give Guy's average speed for the whole 5-hour trip.

   b How far did Guy ride during the 5-hour trip?

   c Calculate his average speed.

Use Map 17 from Exercise 16.03 on page 386 to answer questions **7**, **8** and **9**.

Exercise
16.03

7 What village is located at A11?

Exercise
16.03

8 Approximately how far is it from the traffic lights in A13 along Binstead Way to the lights at the motorway off ramp in D13?

Exercise
16.03

9 Describe a route you could follow to walk from Cannington Place (E9) to Wilmington Ct (F7).

10 The map shows the Mount Ngungun walking trail in the Glasshouse Mountains.

Chris is going to walk from the cark park at *P* to Mount Ngungun.

Mount Ngungun, Glass House Mountains Walking Tracks Info and maps. The image is reproduced with the permission of the Department of National Parks, Sport and Racing, Queensland Government. © State of Queensland.

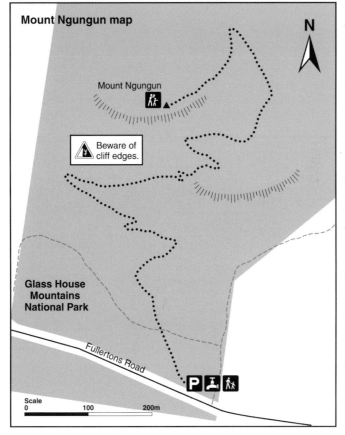

Mount Ngungun map

Mount Ngungun

Beware of cliff edges.

Glass House Mountains National Park

Fullertons Road

Scale
0   100   200m

a In what compass direction will Chris travel for the first part of the walk?

b Use the scale to calculate the length of the walk.

c The walk is quite challenging and Chris will only be able to walk at an average speed of 24 m/min. Approximately how long will the walk to the top take?

d Chris plans to leave the car park at 9 a.m. and walk to the top, spend 40 minutes at the top and then walk back to the car park. At approximately what time will he be back at the car park?

11 Find the shortest distance from *A* to *C* in each diagram.

a

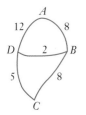

b

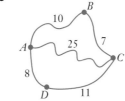

12 Calculate the shortest distance from *A* to *C* on this mud map.

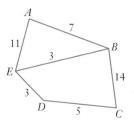

13 The map shows the central region of the Northern Territory that contains Uluru.

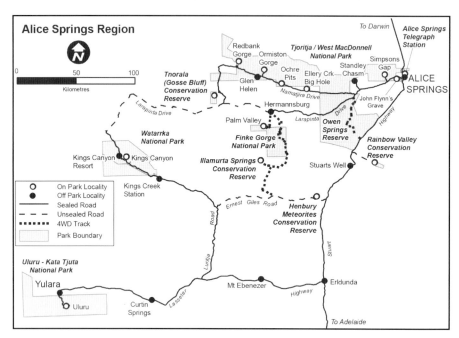

a What type of vehicle can travel to Finke Gorge National Park?

b Jess wants to drive from Alice Springs to Kings Canyon, but she doesn't want to travel on any unsealed roads. Describe a route she could take.

c Describe a route Jess could take to drive from Kings Canyon to Yulara.

d Damian is having coffee at Mt Ebenezer on the way to visit Uluru. Approximately how far is it from Mt Ebenezer to Yulara?

e Damian plans to spend the night in a caravan park on the way to Yulara. Where do you suggest he should stay?

f What compass direction is the trip from Mt Ebenezer to Yulara?

g Tegan is driving from Stuarts Well to Alice Springs. In what compass direction is she heading?

Practice set 4 ●●●●

Section A Multiple-choice questions

For each question, select the correct answer **A**, **B**, **C** or **D**.

1 During exercise, Robbie had an average pulse of 140 beats/minute. How many times did her heart beat during her 40-minute exercise session?

   A   3.5   **B**   210   **C**   5600   **D**   8400

2 Miriam is surveying students about takeaway food. She writes the question 'Describe the frequency with which you consume takeaway food on a weekly basis.' This is NOT a good question because:

   A   the language is too complicated

   B   the meaning of the question is not clear

   C   the question is biased

   D   the question doesn't respect the privacy of the person answering

3 What is the shortest path from P to Q?

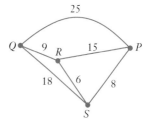

   A P to Q

   B P to S to Q

   C P to R to Q

   D P to S to R to Q

4 Petra earns a gross salary of $63 500. She paid $12 185 in PAYG tax in the last year and pays $17.60 per week for other deductions. Calculate Petra's net weekly pay.

   A $969.23   **B** $986.83   **C** $1938.45   **D** $1973.65

5 Using the information in Exercise 13.01 Question 7, on page 314, which of the following activities will use 3500 kJ?

   A   Sleeping for 12 hours and 30 minutes

   B   Studying for 7 hours and 30 minutes

   C   Bike riding for 2 hours and 55 minutes

   D   Swimming for 1 hour and 35 minutes

6 A map has a scale of 1 : 50 000. Which of the following distances is represented by 64 mm on the map?

   A   0.32 km   **B**   3.2 km   **C**   32 km   **D**   320 km

Exercise 14.02

7 Aaron earns $1538.50 per week. He claims deductions of $715 for travel expenses and $218 for membership of a professional association. He also gives $50 a month to charity. Calculate his taxable income for one year.

    A $80 002   **B** $79 069   **C** $79 019   **D** $78 469

Exercise 16.02

8 A boat travels 24 km in 90 minutes. What is its average speed?

    A   12 km/h   **B**   16 km/h   **C**   18 km/h   **D**   36 km/h

Section B   Short-answer questions

Exercise 13.04

1 Malek is exercising and he takes his pulse. He counts 39 beats in 15 seconds. Calculate his pulse in beats per minute.

Exercise 14.03

2 Rajesh has a taxable income of $91 300. Calculate the 2% Medicare levy payable.

Exercise 15.01

3 Jim's survey asks 'How often do you go out to dinner?' Suggest a series of tick boxes that Jim could provide for people to record their answers.

Exercise 16.03

4 Use Map 17 in Exercise 16.03 on p. 385 to answer these questions.

    a   What feature is at A20?

    b   What are the position coordinates of Careel Reserve?

Exercise 13.03

5 The Thomson family own a 550W TV that they use an average of 7 hours per day. The domestic tariff for electricity is 34c per kilowatt-hour.

    a   How many kWh of electricity does the TV use each day?

    b   How much does the TV cost in electricity each day?

Exercise 16.01

6 Tim walks to school each day. It takes him 20 minutes to walk the 1500 m to school. Find his walking speed in:

    a   metres per minute   **b**   kilometres per hour

Exercise 15.02

7 Joanna's survey included the question:

'Rate your meal at the fabulous Diggies' Steak House:

Great   Yummy   Quite nice'

    a   In what way is this question biased?

    b   Rewrite it so it is not biased.

Exercise 14.04

8 Kane and Anne have kept details of their household expenses from the last year.

Electricity	$3874	Council and water rates	$2613
Insurance	$1297	Internet/phones	$1200

    a   Calculate the total annual cost of Kane and Anne's household bills.

    b   In the coming year they expect increases in all expenses. Increase the total for Kane and Anne's annual household bills by 10% to allow for future price rises. Round up to the nearest dollar.

c Both Kane and Anne are paid monthly. Use your answer to part **b** to calculate the amount they should budget each month to cover household expenses. Round up to the nearest dollar.

d What other items might Kane and Anne have to budget for?

9 Use the map of Tasmania in Exercise 16.02, Question 1 on page 378 to find the compass direction from:

a Miena to Bothwell

b Bruny Island to Hobart

c Dover to Strahan

10 Henri is conducting a survey for a Queensland company. He stands on the corner of an intersection in central Brisbane and asks every 15th person who walks past to complete a short questionnaire.

a In what way is this sample biased?

b How could you ensure that an accurate sample of all Queenslanders is found?

11 Niko has lunch consisting of 2 slices of ham, 1 slice of cheese and 1 slice of buttered bread, followed by a banana.

a Use the table on page 315 to calculate how many kilojoules are in Niko's meal.

b Niko goes swimming in the afternoon. Use the table in Exercise 13.01, Question 7 on page 314 to determine how long she must swim to use the kilojoules from her lunch.

12 Annabel's taxable income for the last financial year is $77 210. She has $334.32 deducted each week in PAYG tax.

a Calculate how much tax Annabel has paid in one year.

b Use this table to calculate how much tax is payable on her taxable income.

Taxable income	Tax on this income
0–$18 200	Nil
$18 201–$37 000	19c for each $1 over $18 200
$37 001–$87 000	$3572 plus 32.5c for each $1 over $37 000
$87 001–$180 000	$19 822 plus 37c for each $1 over $87 000
$180 001 and over	$54 232 plus 45c for each $1 over $180,000

c Does Annabel receive a refund or does she owe money to the tax office? Find the amount she receives or owes.

ANSWERS

Chapter 1

Exercise 1.01

1 a 14 **b** 16 **c** 20 **d** 33
2 a 16 **b** 18 **c** 20 **d** 28
e 45 **f** 17
3 Any combination of two scores that add to 13, e.g. double 6 + 1 or 7 + 6.
4 6 ways: 4 + 3, 6 + 1, 5 + 2, 5 + double 1, 3 + double 2, 1 + double 3
5 a 23
b Double 10 + 3 or double 9 + 5 or double 8 + 7 or double 7 + 9
6 Four throws (4 × 25)
7 3 + double 8
8 To make it harder to get high scores. If the players want to get high scores, they need to place accurate throws.

Exercise 1.02

1 a 46 **b** 90
2 180 (3 × triple 20)
3 Many answers are possible, e.g. 20 + 10 + 7 or 13 + 17 + 7
4 151
5 Rebecca
6 16 or double 8
7 Still 86
8 104
9 a double 17 **b** double 10 + double 7
c 10 + 10 + double 7
10 160
11 a 4 **b** Renata on turn 2
c 58 on turn 3 **d** 2
e 2 throws for combined 10, then double 7

f 29 is an odd number (no number on the board doubles to make 29)
g Teacher to check.

Exercise 1.03

1 Brett
2 −1
3 a 70 strokes **b** 5 strokes
4 a The temperature is 10° below zero.
b −4°C **c** −7°C **d** −5°C
5 a −2 **b** 3 times **c** 376 runs
d Many answers are possible, e.g. batter gets out twice and scores zero on the other 2 balls.

Exercise 1.04

1 a False, $4 + 2 \times 3 = 10$ **b** True
c True **d** True
e False, $2 \times 4^2 = 2 \times 16 = 32$
f False, $48 \div 4 \times 3 = 12 \times 3 = 36$
g False, $20 - 5 + 8 = 23$ **h** True
i True
2 a 3 **b** 3 **c** 22 **d** 36
e 75 **f** 1 **g** 12 **h** 10
i 2 **j** 14 **k** 40 **l** 140
m 150 **n** 26 **o** 30
3 a $(4 + 7) \times 5 = 55$ **b** $60 \div (5 + 7) = 5$
c $(3 \times 2)^2 = 36$ **d** $(6 + 8) \times (9 - 5) = 56$
e $(3 \times 4 + 5) \times 2 = 34$ **f** $(28 - 4 \times 5) \times 2 = 16$
4 When she pressed the ▬ key in the middle, the calculator added 3 and 6 first, before multiplying by 5.
5 a The expression represents the sum of the points for 4 red balls, 2 brown balls, 1 pink ball and 2 black balls.
b 17 points

6 a Do the × first

b Explain the order of operations

c $(48 - 8) \times 3$

Exercise 1.05

1 $6870

2 $39 000

3 $84

4 a $920   **b** $47 840

5 $7.20

6 4.5

7 a 9h 45 min   **b** 113 h 45 min

8 12 minutes

9 Rockets by 2

10 Neither team is in front. The score is 110 each.

11 $1060.50

Exercise 1.06

1 B   **2** B   **3** C   **4** B   **5** B

6 a 1600 m   **b** 4

7 $26

8 a $34   **b** Yes   **c** Yes

9 a Paige knits approximately 2 rows per minute.

b approximately 25 minutes

Exercise 1.07

1 a 16.1   **b** 29.8   **c** 14.6

d 13.3   **e** 104.6   **f** 195.2

2 124.73

3 Many answers are possible, e.g. 16.27 or 16.32

4 Teacher to check.

5 a Mia: wave 1, 4.9; wave 2, 7.3; wave 3, 8.5; wave 4, 7.6

Elissa: wave 1, 7.0; wave 2, 7.8; wave 3, 5.2; wave 4, 8.2

b Mia: 7.6 + 8.5 = 16.1; Elissa: 7.8 + 8.2 = 16.0

c Mia

6 3.9 m

7 538.5 km

8 a 8745 m   **b** 8.7 km

9 a 79.3 is close to 80 and 155.8c is close to $1.50

b $120   **c** $123.55   **d** $3.55

10 a Closer to $150 because $1.35 is closer to $1 than $2.

b $207.63

c If she uses the same amount of LPG each month, her budget allowance isn't enough.

11 $16.21

Exercise 1.08

1 a 1600 envelopes   **b** 3 boxes

c 200, 400, 600, 800   **d** 5 boxes

2 a 1400 mm is not a multiple of 300 mm.

b 1500 mm

c 2400 mm

d 1.2 m

3 600 m

4 a 20 cm of pink, 40 cm of white

b Pink $4.40, white $8.80, total cost $13.20

5 Posts: 8 of 2.7 m and 1 of 2.1 m. Top rails: 8 of 2.4 m.

Exercise 1.09

1 a 24 cents   **b** 24 cm

2 a 90c   **b** 90 cm

3 a 2 h 30 min   **b** 3 h 48 min

c 1 h 24 min   **d** 2 h 54 min

4 a 12   **b** 6   **c** 4 years 6 months

5 2 years 3 months

6 a 9   **b** 5 years 9 months

7 a 6 months 15 days   **b** 8 months 27 days

c 3 months 7 days   **d** 5 months 9 days

8 a 3   **b** 18 overs and 3 balls

c Teacher to discuss.

Keyword activity

1 Double   **2** Triple

3 Bullseye   **4** Integer

5 BIDMAS   **6** Positive

7 Indices   **8** Brackets

9 Negative   **10** Darts

11 EFTPOS

```
X P O S I T I V E Y T
D M I C T N P D L S G
I R A U B I D M A S W
N B N E G A T I V E J
T U R W Z C F R C J A
E L O A K T S X N E I
G L D M C L R H C Q S
E S H Y W K G I V O B
R E S A J R E F P O Y
B Y E H D A R T S L R
K E L P O X F N S R E
D O U B L E M F O Z U
```

Chapter review 1

1 a 9 **b** 13 **c** 22 **d** 23
2 a 34 **b** 84
3 Mike won by 5 strokes.
4 a 18 **b** 14 **c** 4
5 a 174.8 **b** 19.9 **c** 151.3
6 a $96 **b** No, it's too small.
7 $10.60
8 a 16 **b** 16
9 a 28 kg **b** 135 m **c** 5 mm
10 15
11 Any whole number in the range 141 to 144.

Chapter 2

Exercise 2.01

1 a $33 **b** 84 kg
  c 252 cm **d** $307.50
  e 216 marks **f** 595 000 people
  g 152.29 kg **h** 5 m
  i 294 students **j** $312.70
  k 12 L **l** 113.6 ha
2 a 96 cm **b** 84 hours
  c 219 days **d** 3900 mL
  e 23c **f** 98 days
  g 1350 m **h** 600 kg
3 258 students
4 $412.50
5 a $0.81 **b** $2062.50

6 3.72 hours
7 782 seedlings
8 $52.40
9 1275 people
10 11 700 seats
11 24.31 kg
12 337.5 MB

Exercise 2.02

1 a 80% **b** 55%
  c 80% **d** 88.75%
2 a 87.5% **b** No
3 41.8%
4 a 8.3% **b** 18.75% **c** 12.5%
  d 37.5% **e** 52% **f** 11.4%
  g 25% **h** 21.4% **i** 13.7%
  j 13.9% **k** 21.4% **l** 6.4%
5 75%
6 17.6%
7 49.4%
8 55.9%
9 52.5%
10 38.1%
11 a 34.8% **b** 47.8%
12 a English 60%, Mathematics 54.3%, Science 64%
  b Science

Exercise 2.03

1 a 152 kg **b** $2650
  c 157.5 m **d** 13.3 L
2 $9 **3** $682
4 $35.37 **5** $234
6 $20 790
7 a $94 080 **b** $158 080
8 24 hectares **9** $2041.20
10 $184 **11** $563.33
12 a $516.25 **b** $567.88
  c Teacher to check.

Exercise 2.04

1 a $64.50 **b** 1440 L
  c 105.05 kg **d** 714 students
  e 7.2 hours **f** 3 weeks
2 $247.50 **3** 31 141
4 $679.15 **5** 1438
6 a $427.50 **b** $877.50
7 813 students
8 a $39.57 **b** $1939.13
9 $172.01
10 $27 993
11 $1032
12 a i $670 **ii** $549.40 **b** $494.46

Exercise 2.05

1 a 28.6% **b** 29.6% **c** 83.1% **d** 55.8%
2 a 18.6% **b** 40.2% **c** 12% **d** 15.4%
3 87.5%
4 45.5%
5 a Profit, $15 **b** 10%
6 a Loss, $500 **b** 11.8%
  c Keiran might be happy as he used the car for a year.
7 $272.25
8 a $585 **b** $23 985 **c** $21 100
  d Loss **e** 12%
9 a $315 **b** $324 **c** Profit **d** 2.9%
10 45.8%

Exercise 2.06

1 80%
2 38 385
3 a 19.7% **b** 22.4 g
4 $8625 **5** $11 200 **6** 7.6%
7 $703.80 **8** 69.02 g **9** $754.40
10 $350 175 **11** 13.5% **12** $691.25
13 a 0.83% **b** 4:55:30
14 164.5%
15 $4500
16 22%
17 30.8%
18 $41 850

Chapter review 2

1 a $32 **b** 210 mL **c** 11.25 ha
  d 60 cm **e** 3600 g **f** 81 mins
2 a 273 people **b** $945
  c 16 questions
3 a 85% **b** 25% **c** 10%
  d 75% **e** 25% **f** 15%
4 a 3.5% **b** 93% **c** 36.1%
5 a $367.50 **b** 93 kg
  c 18 000 people
6 a $1478.05 **b** $2310
7 a $54 **b** 63.75 m **c** 6 weeks
8 a $45 **b** 69 006 people
9 a 20% profit **b** 40% profit
  c 26.3% loss **d** 9.5% loss
10 a 70.6% **b** 114.3%
11 a 117 females
  b i 30.5% **ii** 2 g
  c $16

Chapter 3

Exercise 3.01

1 $17 is too small. She will earn 14 + 14 + 14, which is more than $17.
2 After he has paid for the T-shirt he will have less on his credit card, not more.
3 65 minutes is more than an hour. If she catches the 8:15 a.m. bus, she will get to work after 9:15 a.m.
4 Stuart spends less than an hour at the gym each day. The answer should be less than 5.
5 $65 is too much. Two 500 g bags make 1 kg. One bag costs less than $20. Two bags will cost less than $40.
6 The answer should be in litres not km.
7 The answer is too much. Interest for each year is less than the amount Dimitri has in his account.
8 10 mm fit inside each cm. There should be a lot more mm than the number of cm.
9 Wrong, the answer should be more than $135.
10 Could be right.
11 Wrong, 550 is more than half of 800. The answer should be more than 50%.

12 Could be right. The answer should be a bit more than 24c.

13 Wrong, her profit was $5, which is less than her cost. The profit is less than 100%.

14 Could be right. The answer should be a bit less than $220 000.

Exercise 3.02

1 a How many days without rain did Danica have in Fiji?

 b It rained on 25% of 12 days

2 How much Mark's car will cost, the price is 5% more than $36 000.

3 How much will Charlotte have to pay? The bill is 5% less than $72.

4 What was the percentage increase? Price went up from $50 to $54.

5 What dates required for Grandma's check? March 12 and 8 days, 6 weeks

6 Does 1 bottle hold enough? 8 mL, 3 times/day, 5 days.

Exercise 3.03

1 a

16 18 20 22 24 26 28 30
April

 b 18th and 24th

 c Draw a diagram, work backwards.

2 a 18 children, $158

 b 15 children, 215: 4 adults and 16 children is correct

 c Guess, check and improve

3 a 16

Day	1	2	3	4	5
Number of treats	32	16	8	4	2

 b total number = 62

 c Make a table or list

4 a 12 m, multiply   **b** 187.4 km

 c Solve a simpler problem, draw a diagram

5 a

Town 80 100 160 200 240 300 320 400 480
      km km   km km km   km km   km   km
        F P   F P F   P F   P, F   F
          P = petrol F = food

 b 400 km   **c** Draw a diagram

6 a $460 000   **b** $10 000, $11 500, $21 500

 c Break it into smaller pieces.

Exercise 3.04

1 share, divide

2 per, multiply

3 deducts, subtract

4 how many more, subtract

5 per, multiply

6 reduce, subtract

7 total, add

8 at this rate, divide

9 groups of, multiply

10 shared equally, multiply

11 doubled, divide by 2

12 discount, addition

Each activity in question **10** to **12** has happened already and we're working backwards.

13 Answers from Exercise 3.02.

 1 percentage of, multiply

 2 percentage of, multiply and increase, add

 3 percentage of, multiply and discount, subtract

 4 increased, subtract

 5 no word clues

 6 per day, multiply

Exercise 3.05

Part A

1 The amount of concentrate and paddock grass required each day. We know
 • the horse weighs 500 kg
 • a horse needs 2% of its weight in food each day
 • the horse requires 70% of its food to be concentrate and 30% to be paddock grass

2 2% of → multiply
 70% of → multiply
 30% of → multiply

3 $0.02 \times 500 = 10$ kg
 0.70×10
 0.30×10

4 7 kg concentrate, 3 kg paddock grass

5 Correct, mostly concentrate and $7 + 3 = 10$

6 Saskia should give her horse 7 kg of concentrated food per day and allow the horse to eat 3 kg of paddock grass each day.

Part B

1 There were 9 fine days.

2 The new price of the car is $37 800.

3 Charlotte will have to pay $68.40.

4 The store owner increased prices by 8%.

5 Grandma's first appointment is on March 20 and her second is on April 23.

6 Yes, there is exactly the required amount in the bottle.

Chapter review 3

1 a Wrong, it's not enough. The cost is about $9 \times $30.

  b Wrong, the price including GST has to be more than the price without GST.

  c Wrong, it's too much. She will save a small part of $380.

  d Could be right.

2 a How much wire will be left.

  b The roll is 20 m long and she will use $8 + 3.6$ m.

3 a The bus leaves on the o'clock. The trip takes 1 h 20 min. It takes 30 min to get to the bus. You have to arrive before 3:45 p.m.

  b What's the latest time you can leave home and get to the game before it starts?

4 a Share, divide   **b** Cheaper, subtract

  c per, multiply   **d** discount, subtract

5 a 12   **b** 2 p.m.

Chapter 4

Exercise 4.01

1 a 50 km/h   **b** 25 words/min

  c 50 L/h   **d** $17.50/kg

  e 7.27 m/s   **f** $26/h

  g 250 g/L   **h** 600 revs/min

2 18 L/m^2

3 1350 L/h

4 a 40 L/container   **b** 4 L/day

  c $5\frac{1}{3}$ g/cm^3 **d** $3.20/DVD

  e 35 mm/day   **f** 14.5 km/L

  g $11.40/m   **h** 256 vibrations/s

  i 40 sheep/ha   **j** $1.58/L

5 9 m^2/L

6 $120/day

7 a $8.50/kg   **b** $12.48/kg   **c** $2.45/kg

  d 55c/roll   **e** $2.45/bottle

Exercise 4.02

1 a 5.6 m/s   **b** 13.9 m/s

  c 22.2 m/s   **d** 30.6 m/s

2 a 4.8 km/h   **b** 1.3 m/s

3 a 45 km/h   **b** 144 km/h   **c** 3.5 km/h

4 40 kg/ha

5 1.8 L/h

6 a 1.2 g/mm   **b** 1.2 t/km

  c the same   **d** 12 g/cm

7 a 3500 L/h   **b** 58.3 L/min   **c** 1.0 L/s

8 a $2.80/L   **b** 0.28c/mL

9 a 10.4 m/s   **b** 37.44 km/h

10 garden pots (76 mL/m^2)

Exercise 4.03

1 a $2.30   **b** $2.20   **c** $2.25

2 a 50 mL: 72c/10 mL; 80 mL: 70c/10 mL

  b 80 mL

3 a 40c and 35c

  b box containing 6 eggs

  c 3 boxes containing 6 eggs

4 350 g, as it's the cheapest per 100 g

5 a 1 kg   **b** 1 kg for $3.50

6 a $145, $140, $139, $142

  b one 2.5 g packet and one 5 g packet

7 Teacher to discuss

8 The 1 L bottle is the better value. The 750 mL bottle is the equivalent of $11.93 per L.

Exercise 4.04

1 a 112   **b** 196   **c** 336

2 $248

3 a 3 h   **b** 1.5 h

  c 4.8 h or 4 h 48 min

4 6.25 kg

5 a 720 kg   **b** 15 trees

6 a 5460 km

  b i 2.0 h   **ii** 3.1 h   **iii** 4.1 h   **iv** 15.2 h

7 a 960   **b** 270   **c** 25 m^2 **d** 95 m^2

8 768 kg

9 a 750 mL   **b** 1500 mL   **c** 1.5 L

   d 36 L   **e** 40 min   **f** 40 h

10 a $8000   **b** 35 weeks   **c** 3 weeks

11 a 18 h   **b** 35 h   **c** 24 h

12 a 21.6 m   **b** $270   **c** 18 m

   d 10 windows

13 a 504 L   **b** $1134   **c** $2232   **d** 15 h

Exercise 4.05

1 a 50.15 L   **b** $80.24

2 13.9 L/100 km

3 820 km

4 9.5 L/100 km

5 a 10.1 L/100 km   **b** 8.8 L/100 km

   c 8.5 L/100 km   **d** 7.3 L/100 km

   e 7.8 L/100 km

6 800 km

7 a 7.15 L   **b** $10.73

8 a 47 L   **b** $82.25

9 a They both use 1.8 L to complete the job.

   b Use whichever mower he wants. However, with the ride-on, there will be less noise pollution.

10 a 0.4 L/h   **b** 3.2 L

    c 7.5 h

Keyword activity

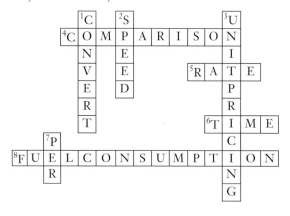

Chapter review 4

1 $64/h

2 20.8 m/s

3 108 000 km/h

4 a i $1.33   **ii** $1.07   **iii** $1.30

   b 750 mL

5 $2310

6 $106

7 Choose $36.50/h. It pays $60 more per week.

8 a White car 8.5 L/100 km, Red car 9.1 L/100 km

   b White car

Chapter 5

Exercise 5.01

1 a 5 cars   **b** 35

   c 6 a.m.–8 a.m.   **d** Teacher to check.

2 a 13   **b** Friday   **c** 48

   d Teacher to check.

3 a 500 000 people or half a million

   b Sydney, 4 900 000

   c Darwin, 100 000

   d The scale means we can only estimate to the nearest 100 000 but the actual population is more exact than that.

   e Adelaide   **f** Teacher to check.

4 a 2%   **b** NSW

   c Victoria and Northern Territory

   d No, it only tells you the percentage increase, not the actual numbers.

   e 512

5 a Peanut butter   **b** Biscuits

   c Cereal   **d** 28%

   e Peanut butter   **f** Biscuits

   g 14 g   **h** Cereal and biscuits

6 a Motor vehicle thefts in the country town are decreasing over these 6 years.

   b 2011   **c** 2013, 55

   d Teacher to check.

7 a 10.5%   **b** 1990 and 1991

   c 1994 and 1995   **d** $20 400

   e i $8640   **ii** $11 760

8 a 15 kWh   **b** 23 kWh

  c Monday, Tuesday, Wednesday in the first week, Friday in the second week

9 a 20°C   **b** 212°F   **c** 32°F

  d 38°C

10 a €10.5   **b** $70   **c** $50

  d Yes, since €25 ≈ $35   **e** €154

11 a 420 kJ

  b 170 cal

  c 2100 kJ

  d about 2100 cal

12 a $120   **b** $200   **c** $280   **d** $40

  e To cover his business costs, e.g. tools.

13 a $24   **b** $48   **c** $60   **d** $60

  e $12

14 a Rugby league   **b** Motor racing

  c Aussie rules and soccer

  d 840   **e** Teacher to check.

15 a Editing

  b Editing, Printing, Binding, Royalty, Transport and Promotion

  c $9.75

16 a Visiting friends

  b 40 people

  c No, on Monday evening there would be more people travelling for work and fewer people travelling for leisure or visiting friends.

Exercise 5.02

1 a 120   **b** People aged over 30

  c Muffins   **d** 36   **e** 80%

2 a **i** 5900   **ii** 16 100   **iii** 11 000

   iv 11 000   **v** 22 000

  b 22 000   **c** $\dfrac{59}{220}$ **d** 73.18%

  e 21.8%

3 a **i** 120   **ii** 95

   iii 108   **iv** 215

  b 215   **c** 64

  d 50.23%   **e** 35.8%

  f More hamburgers, since a large proportion of men prefer them to chicken wraps.

4 a **i** 83   **ii** 187   **iii** 110

   iv 185   **v** 160

  b 24.06%   **c** 59.46%

  d Teacher to check.

5 a **i** 3630   **ii** 9130   **iii** 8920

   iv 3840   **v** 12 760

  b 71.55%   **c** 28.81%   **d** Teacher to check.

6 a **i** 80   **ii** 10   **iii** 75

   iv 45   **v** 120

  b 120   **c** 95   **d** 79.2%   **e** 18.8%

Exercise 5.03

1 a $120   **b** about $23 more

  c about $90   **d** Teacher to check.

2 a 3   **b** about $1750

  c These are for winter, when extra electricity is used for heating and cooking.

  d Teacher to check.

3 a 15 months   **b** approximately 181 MJ

  c June to August, heating in winter, others answers possible.

  d Teacher to check.

4 a 2.05%   **b** 2.30%   **c** $550

  d Money is locked in for the 24 months, you cannot withdraw any of it.

5 a August   **b** January and February

  c January and February, March and December, May and November, or July and September

  d 7   **e** July, August, September

  f Teacher to check.

6 a 34°C   **b** June, July, August

  c A lot of rain from November to March, little rain from April to October.

7 a June, July, August, September

  b about 10°C

  c about 30 mm

  d Teacher to check.

8 Northern hemisphere, because the warmest temperatures are in the middle of the year.

9 a 1% of donations   **b–d** Teacher to check.

  e 9180 L

1

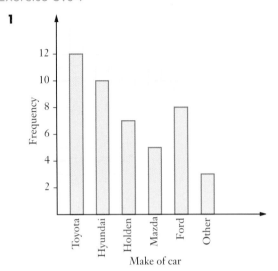

2

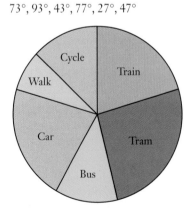

Rainy days in June

= 4 days

3 a 7%   **b** other
  c 31%   **d** 548
4 a 73°, 93°, 43°, 77°, 27°, 47°
  b

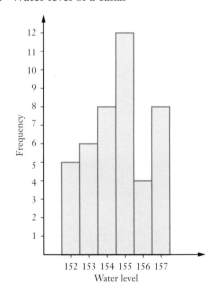

5 a i 70 cm   **ii** 140 cm
  b i 3 years   **ii** 11 years

c 0 and 1 year
d about 2 years 10 months
e fairly flat

6

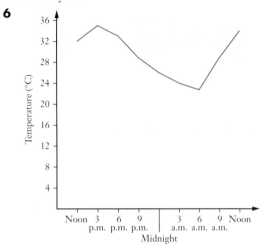

7

Stem	Leaf
3	7 7 8 8
4	1 3 4 4 5 5 6 6 8 9 9
5	1 1 1 2 2 3 3 4 4 5 5 5 8
6	0 2 3 6 6 7
7	0
8	
9	0

8 Water level of a canal

9 a

Test marks	Frequency
6–20	7
21–35	12
36–50	16
51–65	10
66–80	14
Total	59

b 59   **c** 36–50   **d** 19

Exercise 5.06

1 a i 6 kg   **ii** 9.5 kg

b i 8 months **ii** 11 months

c 6 and 7 months

d approximately 6 months

e Teacher to check.

2 a 8%

b 40 to 50 years of age

c 7%

d More likely to die from cancer and heart disease.

e True

f Between 25 and 60.

g Between 18 and 42.

3 a A Melbourne, B Sydney, C Brisbane, D Adelaide, E Perth, F Darwin, G Canberra, H Hobart

b Jan 28°, Feb 27°, March 24.5°, April 19.5°, May 15°, June 12°, July 11°, Aug 13°, Sept 16°, Oct 19°, Nov 23°, Dec 26°.

c March and December

d B June, C July, D May, E June, September

e F

f G, 15° in December

g 33°, F, November

h 0°, G, July

4 a April 22, 24°C

b 15°C   **c** 9°C

d 1 unit = 1°C   **e** 16°C

f April 20, April 21   **g** 16 April

h 5

5 a i Time, 4 hours

    ii Temperature, 0.2°C

b 39.7°C

c Midnight Thursday

d 40.4°C

e 12 hours   **f** 2000 (8 p.m.) Saturday

6

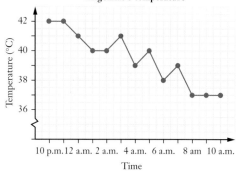

7 a See bottom of page.

b

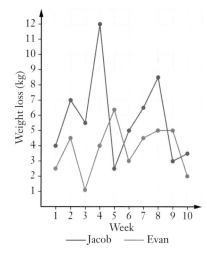

7 a

	Week	Start	1	2	3	4	5	6	7	8	9	10
Jacob	Weight loss (kg)		4	7	5.5	12	2.5	5	6.5	8.5	3	4
	Actual weight (kg)	183	179	172	166.5	154.5	152	147	140.5	132	129	125
Evan	Weight loss (kg)		2.5	4.5	1.1	3.9	6.4	3	4.6	5	5	2
	Actual weight (kg)	137	134.5	130	128.9	125	118.6	115.6	111	106	101	99

c

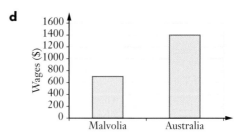

Graph: Actual weight (kg) vs Week, with lines for Jacob and Evan.

d Teacher to check.

e **i** 58 kg   **ii** 38 kg

f Jacob 31.7%, Evan 27.7%

Exercise 5.07

1 a Green 31.5c, Octas 33c, System Two 32.5c

b 1c   **c** the first graph

d The scale on the second graph starts at zero; the scale on the first graph only goes from 30 to 33.

e Green company, to make their calls seem cheap.

2 a Brisbane   **b** Darwin   **c** 144c

d about 14c

e David only used from 135c to 155c and this makes the gaps look bigger.

3 Small: graph the actual population; Big: graph the increase in the population.

4 a 120c   **b** $80

c The cost of petrol is in *cents*, the cost per barrel is in *dollars*: vertical scale is different for each graph

d The two graphs move up and down in a similar fashion: this suggests the two things are related.

5 a Australia's average weekly wages are *much* greater than in Malvolia.

b No vertical scale, using a 3D symbol instead of a column.

c a column

d

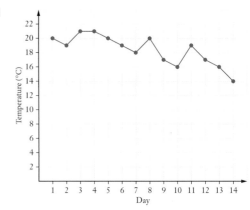

Bar graph: Wages ($) for Malvolia and Australia.

6 Teacher to check.

Chapter review 5

1 a 40   **b** 6 a.m. to 8 a.m.

2 a 2 300 000   **b** 2 200 000

3 a 13%   **b** approximately 1.6%

4 a 42 euros   **b** $84

5 a Motor racing   **b** 595

6 a

	High-speed internet	No high-speed internet	
City	105	38	143
Country	89	42	131
	194	80	274

b 131   **c** 73%   **d** 29%

7 a $120   **b** approximately $52

8 a approximately 32MJ

b December to February

c No heating in the summer months and possibly less cooking.

9 a 3.15%   **b** $1375

10 categorical, pie graph, change, time, it keeps the detail of the data, histogram

11

Graph: Temperature (°C) vs Day.

12

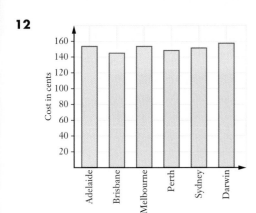

10

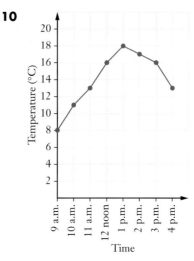

Practice set 1

Section 1

1	B	**2**	C	**3**	C	**4**	A
5	D	**6**	D	**7**	B	**8**	D

Section 2

1 3024 mL

2 a 5 kg   **b** $112 500
    c 192 minutes   **d** 6480 mL

3 a 2011   **b** Inner city   **c** 2011, 150

4 a 79   **b** $\frac{1}{5}$ **c** 29   **d** −2

5 a 92%   **b** 3%   **c** 15%   **d** 25%

6 a 85 m   **b** 17 kg   **c** $10

7 a

Age group	Exercise	No exercise	Total
Under 40	60	50	110
40 and over	65	25	90
Total	125	75	200

   b 37.5%   **c** 52%

8 a 675 km   **b** $64.95

9 $24 592

11 23.31%

12 56.7L/day

Chapter 6

Exercise 6.01

1 a 13 : 25   **b** 17 : 100   **c** 5 : 12
   d 150 : 1   **e** 31 : 51 : 101

2 a 1 : 4   **b** 12 : 1   **c** 2 : 3
   d 3 : 1   **e** 8 : 12 : 1   **f** 1 : 2 : 3

3 a $1
   b i 3 : 2   **ii** 3 : 1   **iii** 2 : 1   **iv** 1 : 3

4 a 27 : 29   **b** 19 : 27   **c** 29 : 75
   d 29 : 19   **e** 27 : 29 : 19

5 a 3 : 250   **b** 23 : 30   **c** 37 : 120
   d 10 : 3   **e** 2 : 5   **f** 300 : 173

6 a black to red   **b** purple to yellow
   c green to grey   **d** green to all
   e green to red to black
   f yellow to purple to grey

7 a 4 : 3   **b** 3 : 4   **c** 3 : 2   **d** 2 : 1
   e 3 : 2   **f** 3 : 1   **g** 3 : 2   **h** 6 : 1
   i 5 : 1   **j** 8 : 7   **k** 7 : 13   **l** 4 : 1
   m 2 : 3 : 1   **n** 2 : 1 : 6   **o** 3 : 5 : 10   **p** 3 : 2 : 4

8 a 1 : 2   **b** 2 : 3   **c** 1 : 3   **d** 4 : 1
   e 2 : 3   **f** 1 : 2   **g** 5 : 3   **h** 3 : 1

9 a $3:20$ **b** $1:20$ **c** $3:1$
d $1:8$ **e** $3:1:20$

10 a $8:3$ **b** $7:3$ **c** $1:4$
d $6:1$ **e** $16:7:1$

11 a $5:8$ **b** $1:13$ **c** $1:2$
d $3:16$ **e** $8:5:3$

12 heavy-duty cleaning

Exercise 6.02

1 a $9:5$ **b** $\dfrac{5}{14}$

2 a $6:2$ **b** $3:1$ **c** $\dfrac{3}{4}$

3 The ratios of red : yellow in the mixtures are
$20:40$ and $4:8$. The ratios are equivalent.
They both simplify to $1:2$.

4 a $5:2000$ **b** $1:400$

5 It looks more green than blue. Possibly Lara got
the ratio around the wrong way. Maybe she made
a $1:4$ ratio instead of $4:1$.

6 a $3:2$ **b** $1:2$ **c** $3:1$ **d** $1:1$

7 72 mL

8 a 5 mL **b** 10 mL **c** 15 mL **d** 20 mL

9 a $\dfrac{1}{4}$ **b** 2 cm **c** $3:1$
d i 2.5 cm **ii** 6.7 cm

10 a 10 **b** 9

11 a zinc 25 g, nickel 5 g
b copper 112 g, zinc 40 g
c 160 g

12 a 5 cups **b** 18 cups

13 a 2000 mL **b** 2 L **c** 150 mL
d 25 mL **e** general perennial weeds

Exercise 6.03

1 56 mL blue, 4 mL black
2 18 mL red, 6 mL yellow
3 8 mL red, 2 mL yellow, 2 mL white
4 20 mL red, 80 mL yellow, 40 mL white

5 400 mL

6 $4200

7 Cement 30 kg, sand 60 kg, gravel 90 kg

8 a Toby $840, Vinson $1260
b Toby $1750, Vinson $350

9 399 boys

10 silver beads 20, porcelain balls 15, crystal
eyedrops 10

11 160 mL

12 a 2 g **b** gold 315 g, copper 84 g

13 a 150 mL **b** 400 mL **c** 200 mL

Exercise 6.04

1 a 1.5 cm **b** 3 cm **c** 4.5 cm
d 7.5 cm **e** 9 cm **f** 3.5 cm
g 3 cm

2 a 1.25 cm **b** 2.5 cm **c** 3.75 cm
d 6.25 cm **e** 7.5 cm **f** 2.5 cm
g 1.9 cm

3

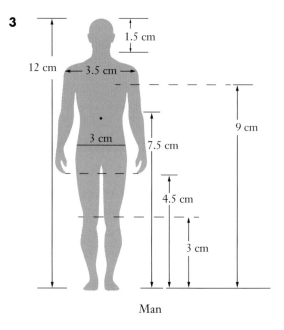

Man

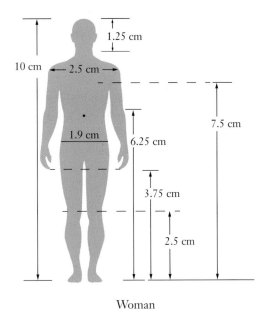

Woman

Exercise 6.05

1 **a** 12 m   **b** 30 m
   c 4.5 m   **d** 20 mm or 2 cm
   e 16.2 m   **f** 208 cm or 2.08 m
2 1100 m = 1.1 km
3 25 m
4 Length 4 m, width 2.5 m
5 **a** 13.5 cm   **b** 720 cm or 7.2 m
   c 16 cm   **d** 62.4 cm
6 4 m
7 6.4 km

Chapter review 6

1 **a** 1 : 3   **b** 75%
2 **a** 2 : 5   **b** 2 : 1   **c** 5 : 2   **d** 3 : 5
3 250 mL
4 3 mL
5 **a** Manal $64, Eddie $16
   b 24 km, 48 km
6 12 mL blue, 8 mL green and 4 mL white
7 48 mm
8 **a** 92 cm   **b** 11.2 cm

Chapter 7

Exercise 7.01

1 **a** $15   **b** 44 000 people   **c** 1.05 kg
   d 810 letters   **e** 750 m²   **f** 48 L
   g 60 066   **h** 6 mins   **i** $21.08
   j 60 people
2 **a** 7 days   **b** 21 days
3 25 299 cars
4 **a** $29   **b** $2.44
5 **a** $9   **b** $234
6 113 students   7 $39.10   8 $899.70
9 287 passengers
10 **a** 324 000 km   **b** 314 160 km
11 **a** 360   **b** 196
12 O+ 9 934 000   A+ 7 698 850   B+ 1 986 800
    AB+ 496 700   O– 2 235 150   A– 1 738 450
    B– 496 700   AB– 248 350

Exercise 7.02

1 **a** 85%   **b** 85%   **c** 40%   **d** 75%
2 **a** English 53.75%, Maths 81.18%,
      Design 58.46%, Science 65%,
      Business Studies 84.29%, Health 75.56%
   b Business Studies, Maths, Health, Science,
      Design, English
3 Tiago 54.5%, Luke 62.2%, Luke is more successful.
4 Muesli bars 31.4%, Jam 49%, Jam has the higher
   percentage of sugar.
5 **a** Colombia 8.78%, Iran 7.07%
   b Colombia
   c Teacher to check.
6 **a** 5%   **b** 37.5%   **c** 1.25%   **d** 25%
7 **a** Monday 28.6%, Wednesday 35%,
      Saturday 30%
   b Wednesday   **c** 31.25%
8 A 25%, B 26.7%, B has the highest percentage of
   alcohol.
9 **a** Clarkson 120.6, Bailden 126.3
   b Bailden
10 Victoria 3%, Western Australia 2.75%, Victoria
    has the higher rate of stamp duty.

Exercise 7.03

1 a $720 **b** $60 **c** $130.20
  d $109.20 **e** $1029 **f** $412.25
  g $34.88 **h** $2056.20
2 a $864 **b** $3264
3 a $202.50 **b** $1702.50
4 a $72 **b** $3.96 **c** $45.83
  d $312.94 **e** $442.74 **f** $17.55
5 $501.88
6 0.39%
7 a 0.7% **b** 0.1615% **c** 4.2%
  d 0.0230% **e** 0.3231% **f** 2.1%
8 a $84 **b** $101.77 **c** $126
  d $31.07 **e** $155.08 **f** $189
9 a $6.69 **b** $371.69

Exercise 7.04

1 a $18 000 **b** $48 000 **c** $800
2 a $261 **b** $1761 **c** $73.38
3 a $3610
  b i $4455 **ii** $58 455 **iii** $2435.63
  c i $150.94 **ii** Yes, she will have $5900.94.
4 $18 750
5 a $18 687.50 **b** $668 687.50
  c $93 687.50
6 a $8795 **b** $6295 **c** $1397.49
  d $7692.49 **e** $213.68
7 a $1620 **b** 4 years
8 5 years
9 a $1200 **b** 10%
10 a 8.5% **b** 6.9% **c** 5.6% **d** 7.2%

Exercise 7.05

1 a $108 **b** $612 **c** $30.60
  d $581.40
  e No, 20% off gives a price of $576.
2 a $53 199.30 **b** $50 539.33
  c No, 35% gives a price of $49 399.35.
3 a 1160 **b** 348 **c** 17.4%
4 a $262.20 **b** 24%
5 a $1552.50 **b** $747.50 **c** 32.5%
6 a $554.80 **b** 24%

7 a $34.63 **b** 17.5%
8 a $640 **b** $704
9 a $704 **b** Neither, they are the same.
10 2786 people
11 $336.08
12 2517

Exercise 7.06

1 a $620.50 **b** $3029.50
2 a $10 400 **b** $26 400 **c** $440
3 a $273 **b** $300.30
  c Teacher to check.
  d $268.13 so probably $269.
4 192 m^2
5 27%
6 a $2700 **b** $2430
  c No, 20% off gives a value of $2400.
7 a $6650 **b** $332.50
8 54.8%
9 a $57 **b** $60.90
10 a $399.90 **b** $3599.10 **c** $4030
  d $4429.90 **e** $430.90 **f** $215.45
  g 5.99%
11 3.49%

Chapter review 7

1 a $414.72 **b** 105 students
2 a i Megan 74.19%, Jasmine 68.75%
    ii Megan
  b i Broncos 77%, Cowboys 79%
    ii Cowboys
3 a $714
  b i $3100.50 **ii** $16 100.50
4 a $127.50 **b** $229.69
5 a 1.0417% **b** 0.2404% **c** 6.25%
6 a $364.60 **b** $286.08 **c** $437.50
7 a $982.80 **b** $7282.80 **c** $202.30
8 a $31 122 **b** $29 565.90
9 a $1871.28 **b** $727.72 **c** 28%
10 54.0%
11 a $614 **b** $2456 **c** $1964.80

Chapter 8

Exercise 8.01

1 a N   **b** C   **c** N   **d** N
   e N   **f** C   **g** N   **h** N

2 a C   **b** C   **c** N   **d** N
   e C   **f** N   **g** N   **h** C

3 Teacher to check.

Exercise 8.02

1 a

Method of travel	Tally	Frequency
W	IIII IIII IIII IIII IIII I	26
R	IIII IIII IIII IIII III	23
B	IIII IIII II	12
C	IIII IIII	10
T	IIII IIII	9
Total		**80**

b Walk   **c** Train

2 a

Age	Tally	Frequency
B	IIII IIII	9
P	IIII IIII	10
I	IIII IIII II	12
O	IIII IIII IIII	14
Total		**45**

b Children older than 8 years.

c 20%

3 a

Make of car	Tally	Frequency
H	IIII IIII	9
F	IIII II	7
T	IIII I	6
M	III	3
S	III	3
O	II	2
Total		**30**

b Holden

c 10%

4

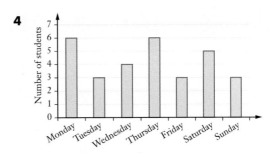

5

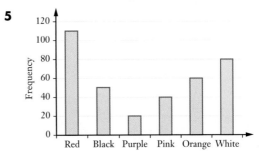

Exercise 8.03

1 a

Number of hamburgers	Tally	Frequency
17	IIII	5
18	IIII	4
19	IIII II	7
20	IIII	4
21	I	1
22		0
23	I	1
24	I	1
25	I	1
26	I	1
27	III	3
28	III	3
Total		**31**

b 16   **c** 23%

2 a

Number of siblings	Tally	Frequency
0	IIII	5
1	IIII IIII	10
2	IIII III	8
3	IIII IIII	9
4	IIII	5
Total		**37**

b 37   **c** 5   **d** 23   **e** 27%

3 a

Age	Tally	Frequency
31–40	IIII IIII	9
41–50	ЖІ ІІІ	8
51–60	ЖІІ І	6
61–70	ЖІ	4
71–80	IIII IIII	8
81–90	ЖІІ	5
Total	/	**40**

b 31–40　　**c** 17　　**d** 32.5%

4

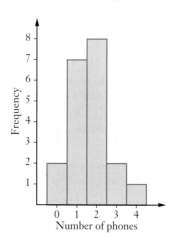

5 a

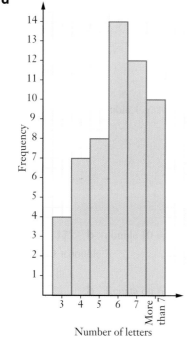

b 11 days　　**c** 6 letters

6 a

Number of phone calls	Tally	Frequency
1	III	3
2	ЖІ	5
3	ЖІ ІІ	7
4	IIII	4
5	ЖІ	5
6	IIII	4
7	II	2
Total		**30**

b

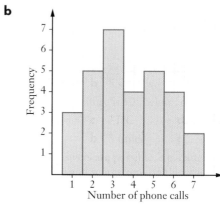

c 5 days　　**d** 3 phone calls

7 a

Heights	Tally	Frequency
150–154	II	2
155–159	IIII	4
160–164	III	3
165–169	ЖІ	5
170–174	ЖІ	5
175–179	IIII	4
180–184	II	2
Total		**25**

b 25

c

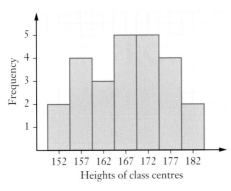

Heights of class centres

d 11 students

e 165–169 and 170–174

Exercise 8.04

1 a

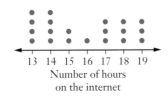

Number of hours on the internet

b 20   **c** 4   **d** 9

2 a

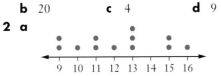

b 3   **c** 25%   **d** 5

3 a

b 12   **c** $28   **d** 5

4 a

Stem	Leaf
7	6
8	1 6 8
9	5 7 8
10	1 5 5
11	2 2 4 7
12	4

b 53%

5

Stem	Leaf
1	1 2 2 3 3 4 6 7 7 9
2	0 0 0 2 3 3 4 5 5 5 6 7
3	0 1 3 3 3 4 5 9
4	1 2 8 8
5	5 5

b 8   **c** 22%

6 a

Stem	Leaf
0	0 0 5 8 9
1	1 4 5 6 7
2	1 2 4 5 6 7 9
3	1 3 5
4	2

b 42 hours   **c** 5   **d** 11

e 21   **f** 52%

7 a stem-and-leaf plot   **b** dot plot

   c histogram   **d** stem-and-leaf plot

   e dot plot   **f** dot plot

   g histogram or stem-and-leaf plot

   h histogram   **i** dot plot   **j** histogram

Exercise 8.05

1 a 28   **b** 1   **c** 12   **d** 94

2 0, 2, 16

3 a 31 45 45 45 49 49 50 50 52 52 55 55 56 58 58
    59 59 60 60 75

   b 31, 75

4 a 30, 47, 48, 48, 49, 54, 59, 59, 63, 64, 68, 68,
    68, 80

   b 30, 80

5 a

Stem	Leaf
2	5
16	1 5 7 7 9 9
17	0 0 1 3 5 6 6 9
18	0 0 2 2 4 5 6
19	7 8
23	0

   b 25 cm – definitely a wrongly recorded
measurement, no Year 11 student would be
this short.

   c 230 cm – probably a wrongly recorded
measurement, a Year 11 student is unlikely to
be this tall.

6 a $1 800 000

  b Reasonable – this could be paid for a particularly large/luxurious house.

7 a $245 000

  b Reasonable – this could be the salary of the CEO or overall business manager.

8 a 16

  b 10 cm, 11 cm, 15 cm, 40 cm

  c All are likely to be wrongly recorded measurements – too small (10, 11, 15) or too large (40) for a foot measurement.

9 a $35 000

  b Reasonable – often a business only makes a small profit in the first year.

Keyword activity

categorical data	C
numerical data	F
frequency table	A
histogram	E
dot plot	B
stem-and-leaf plot	G
outlier	D

Chapter review 8

1 a C   **b** N   **c** N

  d C   **e** N   **f** C

2 a

Colour	Tally	Frequency
Blue	IIII IIII IIII	14
Green	IIII IIII II	12
Red	IIII III	8
Purple	IIII IIII	10
Yellow	IIII IIII	9
Magenta	IIII II	7
Total		60

  b Blue   **c** 20%

d

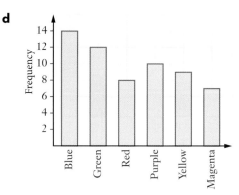

3 a

Temperature	Tally	Frequency
15	II	2
16		0
17	I	1
18	II	2
19	III	3
20	III	3
21	IIII I	6
22	IIII	4
Total		21

  b 21 days   **c** 8 days

d

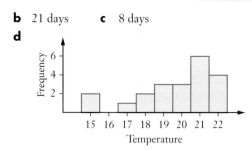

4 a

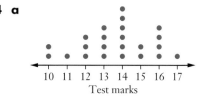

  b 23   **c** 7

5 a

Stem	Leaf
2	5 6 8 8 9 9
3	2 3 3 5 5 6 9
4	1 7 8
5	0 0 2 5 6

  b 21   **c** 56   **d** 42.9%

6 a 19   **b** 7   **c** 43.9   **d** 76

7 a 56   89   93   98   99   100   100   101
104   125

b 56 and 125   **c** Teacher to check.

Chapter 9

Exercise 9.01

1 a

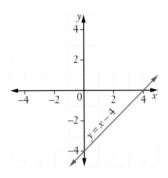

b

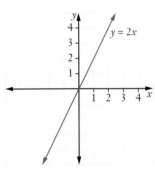

c

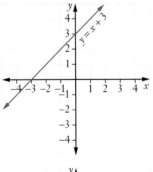

d

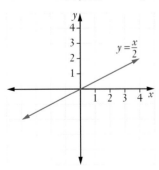

e

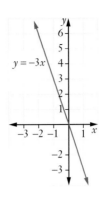

f

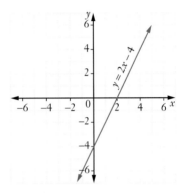

g

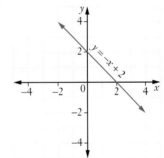

h

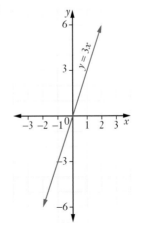

i

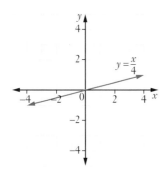

$y = \dfrac{x}{4}$

2

	Gradient	y-intercept
a	$\dfrac{1}{2}$	2
b	3	−1
c	1	6
d	$\dfrac{1}{2}$	1

3 a i 3   **ii** 2
b i −2   **ii** 3
c i −1   **ii** 7
d i 4   **ii** 0
e i $\dfrac{1}{4}$ **ii** −2
f i $-\dfrac{1}{3}$ **ii** −4

4 a positive, 1   **b** positive, 2
   c positive, 1   **d** positive, $\dfrac{1}{2}$
   e negative, −3   **f** positive, 2
   g negative, −1   **h** positive, 3
   i positive, $\dfrac{1}{4}$

Exercise 9.02

1 a 40
   b The number of chirps when the temperature is zero.
   c 8
   d The additional number of chirps per minute when the temperature increases by 1°C.
   e 296   **f** 15°C
   g 840, not realistic
2 a 0   **b** −3.5
   c Number of degrees cooler at which water boils for every extra km in altitude.
   d about 90°C

3 a Laura 14, Bevan 8, Jacob 3
   b speed
   c Jacob, he has the smallest gradient, or he didn't travel as far as the others in 1 hour.
   d Laura bike, Bevan jog, Jacob walk
4 No, the graph says the head circumference should be about 35 cm, which it is.
5 a $40   **b** $\dfrac{1}{4}$
   c Additional cost per ticket
   d $165   **e** $D = \dfrac{1}{2}n + 110$
6 a $50   **b** $1
   c $C = n + 50$
7 a 600 m   **b** 5
   c She starts at 600 and is going down 5 m every second (−5).
   d

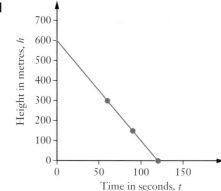

8 a

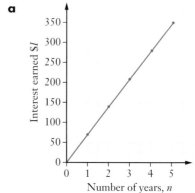

   b 70, amount of interest each year
   c 0, amount of interest at the start
   d $490   **e** $2140

9 a

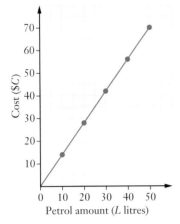

b 1.4   **c** $105

10 a Teacher to check   **b** 2.5 m

c Yes, 4.2 m is more than 4 yards.

Exercise 9.03

1 a Teacher to check

b

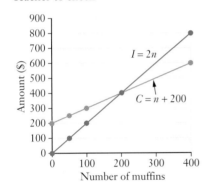

c (200, 400)   **d** break-even point

e No, he needs to sell more than 200 to break even.

f $30

2 a B   **b** 3 min   **c** B

3 a truck 72 km/h, car 96 km/h

b 2   **c** 2 seconds

d car: $m = -96 - 24t$, truck: $m = 72 - 12t$

e Both vehicles were slowing down.

f The truck was heavier.

4 a

Bunches of flowers sold	0	10	20	30	40	50	60
Income received ($)	0	100	200	300	400	500	600

b

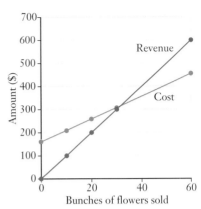

The graphs intersect at (32, 320) which is the breakeven point.

c More than 32 bunches

5 a Income   **b** Expenses

c

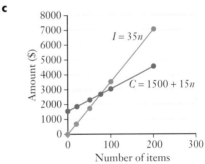

d 75   **e** $500

6 a $60   **b** 10   **c** $6

d It's the weekly cost of running the dog sleds, such as feeding the dogs. It doesn't change with the number of dog sled rides.

7 a $C = 10n + 160$ **b** $I = 20n$

c

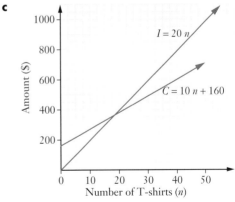

d 16   **e** $740

---

8 a $40   **b** $10

c Monthly service charge

d 20c   **e** $D = 0.2n + 10$

f 150   **g** plan A

Exercise 9.04

1 a 800 m **b** 600 m **c** 200 m **d** 360 m

2 5.5 m

3 a 45 m **b** 12 m **c** 100 m

4 a 33 m **b** 80 m **c** 72 m

5 a

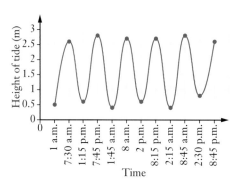

b Approximately 0.9 m

c Approximately 4.30 p.m.

6 a Time

b

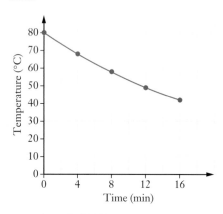

c approximately 74°C

d approximately 10 minutes

7 a

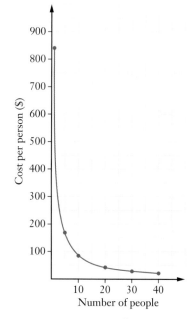

b They both begin and end at about the same place. Other answers possible.

c Anchorage starts low, goes up and comes down again – Perth is the opposite. Other answers possible.

d March, September

e Anchorage is further north of the Equator than Perth, which is south of the Equator.

8 a

Number of people	1	5	10	20	30	40
Cost per person ($)	840	168	84	42	28	21

b

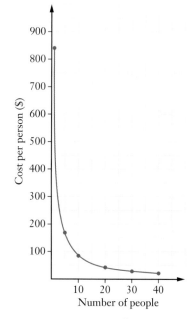

c approximately 30 **d** $30

Exercise 9.05

1 a C **b** A **c** B

2
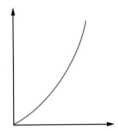

3
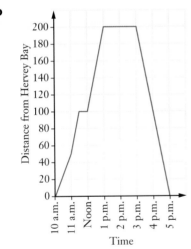

4–7 Teacher to check

8 a Teacher to check

b

9 Teacher to check

Keyword activity

1 algebraic **2** straight
3 gradient **4** y-intercept
5 model **6** intersecting
7 point **8** income
9 costs **10** expenses
11 break-even **12** profit
13 loss **14** curved

Chapter review 9

1 a
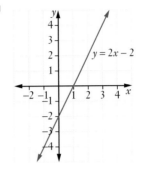

b
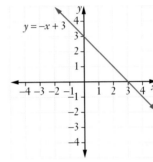

2 a i 1 **ii** −3
  b i −2 **ii** 1
3 a i 4 **ii** −3
  b i −1 **ii** 2
  c i $\frac{1}{3}$ **ii** 0
4 a 3 min **b** 2
  c Time to pack an item
  d 6 items **e** 33 minutes
5 a
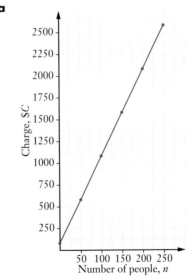

b $1780   **c** 75

d 80, the fixed charge

e 10, the cost per person

6 a $C = 5n + 126$ **b** $I = 12n$

c

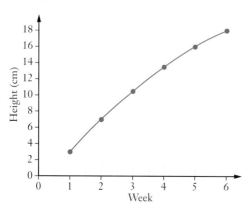

d 18   **e** $364

7 a approximately 48 m   **b** 80 km/h

c 43 m

8

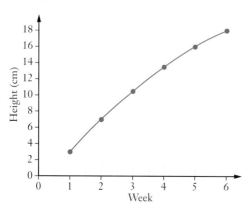

9–10 Teacher to check

Practice set 2

Section 1

1 D	**2** C	**3** A	**4** D
5 B	**6** B	**7** A	**8** C

Section 2

1 a $35   **b** $3535

2 7, 9

3 a

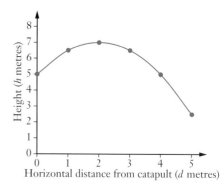

b 7 m

4 75 white tiles

5 a

b 9   **c** 45%

6 a $3680   **b** $4048   **c** $2833.60

7 a

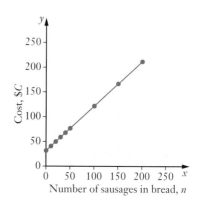

b approximately $185

c 32, this represents the fixed cost

d 0.9, this represents the cost per sausage in bread

e Profit: income $150, costs approximately $100

8 2500 mL

9 a

Waiting time	Tally	Frequency
1–5	II	2
6–10	ИН ИН ИН ИН	20
11–15	ИН ИН ИН III	18
16–20		0
21–25	II	2

b 18   **c** 4.8%

d

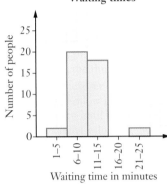

Waiting times

10 Teacher to check

11 172 cm

12 a $6760   **b** $32 760   **c** $546

Chapter 10

Exercise 10.01

1 a $1330   **b** $2660   **c** $69 160

2 a $6875   **b** $3173.08   **c** $1586.54

3 a $67 600

   b A month is usually longer than 4 weeks.

   c $5633.33   **d** $37.14

4 a $249.76   **b** 4.5 hours

5 a $879.23   **b** $1170.43

6 $15.68

7 $4160

8 a $178   **b** 11   **c** Teacher to discuss.

9 a Job 3, the salary   **b** Teacher to discuss.

Investigation: Wages by spreadsheet

1 $616

2 a Total number of hours worked

   b =E6*F17

3 Teacher to check.

Exercise 10.02

1 a $25.80   **b** $34.40   **c** $21.54   **d** $28.72

   e $36.90   **f** $49.20   **g** $46.88   **h** $62.50

2 $123   **3** $122.40   **4** $272.50

5 a 2   **b** $1060.20

6 a i $19   **ii** $28.50   **iii** $38

     iv $399   **v** $114   **vi** $228

   b $741

7 a $18   **b** $81

8 $105

9 a $135.20   **b** 7:30 a.m.   **c** 1 hour

   d 3:30 p.m.   **e** $101.40   **f** 11 a.m.

Exercise 10.03

1 a $962.15   **b** $2290.88

2 $661.25   **3** $1002   **4** $1113.30

5 $791.21   **6** $576.25

7 a i 8   **ii** 7   **iii** 4

     iv 5   **v** $12.32   **vi** $18.48

     vii $18.48   **viii** $24.39   **ix** $98.56

      x $129.36   **xi** $73.92   **xii** $121.95

   b $423.79

Exercise 10.04

1 a $61.25   **b** $221.2   **c** $581.88

   d $1206.63

2 a $486.50   **b** $2655.50

3 $3492.10

4 a $60 644.40   **b** $4664.95   **c** $816.37

5 $6536.62   **6** $5963   **7** $376

8 a $363.69   **b** $5420.89

9 $448

10 Teacher to check.

Exercise 10.05

1 a $2250   **b** $40   **c** $6000

   d $12 500   **e** $6570   **f** $570

2 $117.25   **3** $4603.50   **4** $554.40   **5** $10 600

6 a $18   **b** $61   **c** $159.50

7 a $18   **b** $48.91   **c** $214

8 $2235

9 More than $8.35, $8.50 or $9 would be suitable.

10 $88

11 $412 700

12 a $20 440   **b** $43 400

13 a $16.80   **b** $62.40   **c** 417

14 $2875

15 a $204   **b** $945.20   **c** $2363

Exercise 10.06

1 $1368

2 a $2339.38   **b** $2462.50

3 a She didn't work 30 hours per week.

   b $51.98   **c** $3.75

4 a $31 360   **b** $2613.33   **c** $603.08

5 a $1069.23   **b** $9675

   c The Government wants only a small amount or nothing left in the fund when the person dies.

Keyword activity

1 J   **2** F   **3** G   **4** I

5 K   **6** L   **7** A   **8** C

9 N   **10** E   **11** H   **12** M

13 D   **14** B

Chapter review 10

1 a $656.25   **b** $1312.50   **c** $34 125

2 a $8000   **b** $3692.30

3 a $30.36   **b** $283.36

4 $21   **5** $115.85   **6** $875

7 $6227.50   **8** $40 500   **9** $782

10 $1594.23

Chapter 11

Exercise 11.01

1 a 360 min   **b** 130 weeks

   c 8 min   **d** 10 years

   e 42 days   **f** 12 fortnights

   g 60 months   **h** 42 days

   i 126 days   **j** $6\frac{1}{2}$ years

2 195 min   **3** 15 weeks

4 $1520   **5** 180

6 a $67 020   **b** $1288.85

7 a 3 h 32 min   **b** 9 min 21 s

   c 5 min 30 s   **d** 2 h 15 min

   e 6 h 49 min   **f** 12 min 47 s

8 a 1400 s   **b** 23 min 20 s

   c 5:24 a.m.

9 26 h 45 min

10 10 h 50 min

11 a Teacher to check.   **b** 3 h 26 min

Exercise 11.02

1 a 1144   **b** 1835

   c 0251   **d** 2154

2 a 8:45 a.m.   **b** 1:20 p.m.

   c 11:31 p.m.   **d** 10:45 a.m.

3 a 11:23 a.m.   **b** 5:54 p.m.

   c 1:16 p.m.

4 a 5:20 a.m.   **b** 10:05 a.m.

   c 2:45 a.m.

5 a 3:15 p.m.   **b** 8:31 p.m.

   c 11:55 p.m.

6 a   **b**

   c

7 a 7:15 a.m.   **b** 5:30 p.m.

   c 9:50 p.m.   **d** 11:20 a.m.

   e 0500   **f** 1410

   g 1718

8 a 2030   **b** 2205

9 5:30 p.m.

10 a 8 hours   **b** 3 hours 35 minutes

   c 0305

Exercise 11.03

1 a 5 h 45 min   **b** 5 h 44 min
   c 2 h 20 min   **d** 3 h 53 min

2 a 3 h   **b** 10 h
   c 5 h 15 min   **d** 8 h 10 min

3 9 h 30 min

4 6 h 25 min

5 a 8:00 p.m.   **b** 7:43 p.m.
   c 1:17 p.m.   **d** 2:24 a.m.

6 a 2:15 p.m.   **b** 2:35 a.m.
   c 11:25 a.m.   **d** 7:50 a.m.

7 3:15 p.m.

8 7:45 p.m.

9 a 3 years 8 months   **b** 4 years 1 month
   c Teacher to check.

10 1 year 4 months

11 August 2014

12–13 Teacher to check.

Exercise 11.04

1

	NT	
AWST	ACST	AEST
$-1\frac{1}{2}$	0	$+\frac{1}{2}$

2 a ahead   **b** ahead   **c** same
   d ahead   **e** behind   **f** ahead
   g ahead   **h** same   **i** ahead
   j behind

3 a 11 a.m.   **b** 9 a.m.   **c** 10:30 a.m.
   d 11 a.m.   **e** 11 a.m.   **f** 9 a.m.

4 a 11 p.m.   **b** 10:30 p.m.   **c** 9 p.m.
   d 10:30 p.m.   **e** 11 p.m.   **f** 11 p.m.

5 1:45 p.m.

6 a 5:30 p.m.   **b** 7 p.m.

7 8:30 p.m.   **8** 10 p.m.

9 a–c Teacher to check.   **d** 3:30 p.m.

10 Teacher to check.

11 WA (Perth)

12 Perth, Adelaide (on Summer time)

Exercise 11.05

1 7:50 a.m.   **2** 5:15 p.m.   **3** 8:10 a.m.

4 a 4:52 p.m.   **b** 5:08 p.m.   **c** 5:22 p.m.

5 5:22 p.m.   **6** 8:45 a.m.

7 a 47 min longer   **b** Teacher to check.

8 a, b Teacher to check.

9 20 h 30 min

10 4, 9:45 p.m., 5:15 a.m. (next day), 8:20 a.m., 11:45 a.m.

11 a 6 h 25 min   **b** 11 h

12 a approximately Bulgunnia Turnoff
   b Teacher to check.

13 a 25 h 25 min   **b** Longer
   c Teacher to check.

14 44 min

15 a 11:02 a.m.   **b** 34 min
   c 58 min   **d** Teacher to check.

16 grey is a.m. times, white is p.m. times

17 a 2 buses, Bus 1 can do the 10:30 a.m. departure and so on.
   b 1 p.m., first bus after 11:42 a.m. is 12:05 p.m.
   c Teacher to check.   **d** Teacher to check.

Exercise 11.06

1 a 3:50 a.m.   **b** 3:43 a.m.   **c** 3:52 a.m.

2 a 4:15 a.m.   **b** 4:37 a.m.   **c** 4:05 a.m.

3 a 12 h 49 min   **b** 0.62 m

4 a 5:48 a.m., 6:16 p.m.   **b** 12 h 28 min

5 Mon 14, Wed 23, Thu 24, Sat 26

6 a Sat Oct 26   **b** It is getting earlier.

7 a Between 3 a.m. and 9 a.m. or between 3:30 p.m. and 9 p.m.
   b Between 4:30 a.m. and 10 a.m. or between 5 p.m. and 10 p.m.

8 2:01 p.m.

9 a 12 h 39 min   **b** They are increasing.

10 Teacher to check.

Keyword activity

12-hour time – D

24-hour time – B

timeline – C

time zone – E

timetable – A

Chapter review 11

1 a 12   **b** 49   **c** 54   **d** 8
2 a 1877 seconds   **b** 4 days
3 a i 0417   **ii** 1525   **iii** 1000
   b i 6:15 a.m. **ii** 4:40 p.m. **iii** 11 p.m.
4 a 11 hours, 35 minutes
   b i 7 hours   **ii** 2 hours, 35 minutes
     iii 0113
5 a 4 hours, 25 minutes   **b** 11:25 a.m.
6 a 4 hours, 50 minutes   **b** 10:40 p.m.
   c March 2015
7 a Teacher to check.   **b** 1:30 p.m.
   c 2:45 p.m.
8 a 8:50 a.m.   **b** 8:27 a.m.   **c** 5:01 p.m.
9 a 8 hours   **b** 8:20 a.m.
   c Teacher to check.
10 a 49 minutes   **b** 11:15 a.m.
11 a 2:35 a.m.   **b** 12 hours, 37 minutes
   c 12 hours 16 minutes   **d** 0.24 m

Chapter 12

Exercise 12.01

1 Teacher to check
2 a sample   **b** census   **c** sample
   d sample   **e** sample   **f** census
   g census   **h** sample   **i** sample
   j census   **k** census   **l** sample
   m census   **n** sample
3 a students in Qld schools
   b people of voting age in Qld
   c listeners of a radio station, or readers of a magazine, or musicians
   d students at my school
   e people of Qld
   f the general population
   g Year 12 students
   h rich people
   i people who shop at supermarkets
   j the Australian cricket team
   k internet banking customers of the bank
   l customers of the communications company
   m the population of the town and surrounding areas

Exercise 12.02

1 Teacher to check
2 a CD   **b** DD   **c** AD   **d** PQ
   e WR   **f** OD   **g** AD   **h** CD
   i PQ   **j** DD   **k** WR   **l** OD
3–5 Teacher to check

Exercise 12.03

1 a systematic   **b** random
   c self-selected   **d** stratified
   e self-selected   **f** random
   g systematic   **h** self-selected
   i random   **j** stratified
   k systematic   **l** self-selected
   m random   **n** stratified
   o systematic   **p** stratified
2–4 Teacher to check

Exercise 12.04

1–3 Teacher to check
4 a 81   **b** 57   **c** 24
5 a 33   **b** 19
6 a 99   **b** 47   **c** 52
7 a 48   **b** 27
8 a No. He only surveys people who work in the centre of Brisbane. He misses country people and people who work elsewhere.
   b Teacher to check

Keyword activity

1 J   **2** H   **3** D   **4** F   **5** B
6 A   **7** I   **8** C   **9** E   **10** G

Chapter review 12

1 Sample, too expensive and time-consuming to do a census
2 a i families in the street   **ii** census
   b i companies operating in Qld
     ii sample
   c i voting population of Australia
     ii sample
   d i school students of Qld   **ii** sample
3–4 Teacher to check

5 a 75　　**b** 59　　**c** 16

Practice set 3

Section 1

1 A　　**2** C　　**3** D　　**4** B
5 A　　**6** D　　**7** B　　**8** C

Section 2

1 a 28　　**b** 7　　**c** 4.5　　**d** 11
2 Teacher to check
3 a $4192.31　　**b** $1172.50
4 a 9:30 p.m.　　**b** 12:15 a.m. the next day
5 Teacher to check
6 $23/h
7 a 8:09 a.m.
　　b 7:18 a.m., 7:48 a.m., 8:18 a.m.
　　c 5:39 p.m.
8 a $3880　　**b** $679　　**c** $4559
9 Teacher to check
10 a i 3:30 p.m.　　**ii** 2 p.m.
　　b i 12 midday　　**ii** 10 a.m.
11 a 84　　**b** 45　　**c** 13
12 $8991

Chapter 13

Exercise 13.01

1 a 14 800 kJ　　**b** 22%
2 a 8000 kJ　　**b** 2400 kJ
3 a 5000 kJ　　**b** 1000 kJ
4 a 2100　　**b** 86　　**c** 286
　　d 6　　**e** 586　　**f** 2571
5 a 2310 kJ　　**b** 46 cal
6 a boys　　**b** 2 to 3 years　　**c** 3350 kJ
　　d 4 to 8 years　　**e** 56 000 kJ
　　f No, 1100 kJ too many.
　　g increase
7 a 2000 kJ　　**b** 2 hours
　　c i 10 500 kJ　　**ii** 14 000 kJ
　　　iii no (activities expend 14 000 kJ)
　　　iv Eat more or exercise less.

Exercise 13.02

1 a 2659 kJ　　**b** 1463.5 kJ
　　c 5691 kJ　　**d** 725 kJ
2 a Breakfast 4491 kJ, lunch 4335 kJ, dinner
　　7783 kJ, daily total = 16 609 kJ.
　　b Unless he is very, very active, he will put on
　　weight.
3 Teacher to check.
4 a 3450 kJ　　**b** 115 minutes
5 a 690 kJ　　**b** 765 kJ　　**c** 1020 kJ
　　d 3975 kJ　　**e** 1920 kJ　　**f** 8050 kJ
6 a 306 kJ　　**b** 888 kJ　　**c** 1152 kJ
　　d 14 754.6 kJ　　**e** 76 kg

Exercise 13.03

1

	Power rating (W)	kWh per day	Daily cost (cents)	Monthly cost ($)
a	800	19.2	653	195.90
b	2400	3.6	122	36.60
c	900	0.45	15	4.50
d	1100	1.1	37	11.10
e	100	0.6	20	6.00
f	380	0.19	6	1.80
g	1500	1.5	51	15.30
h	1500	0.375	13	3.90
i	1900	1.9	65	19.50
j	650	0.325	11	3.30
k	950	0.2375	8	2.40
l	40	0.2	7	2.10
m	1400	0.7	24	7.20
n	550	3.3	112	33.60
o	950	0.7125	24	7.20

2–4 Teacher to check.

Exercise 13.04

1 a $18　　**b** $338　　**c** 22
2 a $18, $16　　**b** Teacher to discuss
3 a 75 beats/min **b** 64 beats/min
　　c 130 beats/min
4 a 80　　**b** 100　　**c** 120
　　d 140　　**e** 160　　**f** 180
　　g 200

5 a 172 beats/min   **b** 108 beats/min

c She's working too hard and should slow down.

6 28 breaths/min

7 a $\dfrac{1}{12}$ **b** 1.67 km

8 a 14 min   **b** 50

9 a 255 watts   **b** 3.315 kilowatts   **c** $1.79

10 a 26.4 kilowatts   **b** 9609.6 kilowatts

c $5189.18

Keyword activity

1 d   **2** h   **3** f   **4** b   **5** g

6 a   **7** c   **8** e

Chapter review 13

1 a 1050   **b** 1370

2 a 500 kJ   **b** 175 minutes = 2h 55 min

3 a 1100

b Many answers possible, teacher to check

4 4790 kJ   **5** 3600 kJ

6 0.5 kWh   **7** 8.4 c

8 66 beats/min   **9** Approx. 24c

Chapter 14

Exercise 14.01

1 a $595.50   **b** $4394

2 a $688   **b** $38   **c** $1175

d $1121   **e** $311

3 a $617.75   **b** $552.75

4 $186.62

5 a $674.88   **b** $2402.04

6 a $112.40   **b** $574.14

7 =(D2-D3)*26

8 $26 936

Exercise 14.02

1 a $70 880   **b** $69 920

2 $415

3 a $105   **b** Refund of $669.50

4 a $2035   **b** $52 910

c He will get a bill because he paid $260 × 26 = $6760 in PAYG tax, which is less than the $7897.75 tax payable.

5 a $21 112   **b** $20 887

c $2444   **d** $194 refund

6 She will have to pay $54 more.

7 $77 295.20

8 a $85 020   **b** $25 801.36

c $84 821

d Yes, because she paid more tax ($25 801.36) than was required ($19 113.84).

Exercise 14.03

1

	Income tax	Medicare levy	Total tax
a	$6237	$904	$7141
b	$3564.40	$739.20	$4303.60
c	$7675.13	$992.50	$8667.63
d	$32 957	$2450	$35 407
e	$56 932	$3720	$60 652

2 a $38 710   **b** $4127.75   **c** $774.20

d No   **e** $381.95

3 a $30 274.50   **b** $3745.63   **c** $1440.63

4 a $520   **b** $32

5 Wrong! Australian income tax is progressive. The more your taxable income, the greater proportion you pay in tax. Jacob's tax is $7797 and Garth's tax is $43 132 which is more than 5.5 times Jacob's tax.

6 a Tax = $16 637, Medicare levy = $1544, total $18 181

b $7881

7 a $2700   **b** $900   **c** $65.40

8 $64 250

Exercise 14.04

1 a $5800   **b** $111.54

2 a

Income		Expenses	
Office	$620	Rent	$280
Club	$215	Food	$60
		Mobile phone	$20
		Travel	$60
		Total fixed expenses	$420
		Entertainment	$60
		Savings	$315
		Clothes	$40
Total income	$835	Total expenses	$835

Other answers possible for entertainment, savings, clothes.

b

Income		Expenses	
Office	$620	Rent	$280
Club	$215	Food	$60
		Mobile phone	$20
		Car	$175
		Entertainment	$50
		Savings	$210
		Clothes	$40
Total income	$835	Total expenses	$835

3 a $312   **b** $41 080   **c** $16 224
4 a Income = $420, Expenses = $300
   b $1440   **c** Teacher to check
5 a $158   **b** $223
   c Approx. 113 weeks
   d Teacher to check

Keyword activity

Chapter review 14

1 a $702   **b** $604
2 $89 585
3 Taxable income = $64 600,
   Medicare levy = $1292
4 a $18 359.50   **b** $1577

5 a She should budget. If she saves $80 per fortnight and puts it into a special account she will have a little more than she needed this year to pay these expenses.

6 a He hasn't included rent, electricity, gas, travel to work costs. Possibly he should also include health insurance and other insurances, and a budget for car registration and repair costs. In addition, the items add to more than his fortnightly income.

   b He needs to rethink his finances. He can't afford his expenses on his income.

Chapter 15

Exercise 15.01

1 3
2 4
3 Some people would have nowhere to tick, for example, a 30-year-old.
   ☐ 30 years or under
   ☐ From 31 to 40 years
   ☐ From 41 to 50 years
   ☐ Over 50 years
   Other answers possible.
4 Which of the following do you attend?
   ☐ Concerts   ☐ Special events
   ☐ Films   ☐ Trips away
   ☐ Fundraisers   ☐ Other
   Other answers possible.
5 ☐ Weekly
   ☐ Once a month
   ☐ 2–3 times per month
   ☐ Over 3 times per month
   Other answers possible.
6 For the activities you attend, the cost is:
   ☐ cheap
   ☐ OK
   ☐ expensive
   ☐ I don't attend activities because they are too expensive
   Other answers possible.
7 Teacher to check
8 A box could be provided. Other answers possible.
9 Teacher to check

Exercise 15.02

1 a use of 'fascinating' and 'wet'
  b use of 'disgusting'
  c no negative options
  d use of 'exciting' and 'boring'
  e second half of question shouldn't be used
  f no positive options
  g use of 'boring'
  h use of 'greatest', yes/no response only
  i use of 'one of those'
  j no negative options
2 Teacher to check

Exercise 15.03, 15.04

Teacher to check

Chapter review 15

1 a knowing what time frame to consider, other answers possible
  b Make the question 'How often do you drink each week?'
  ☐ Once a week
  ☐ 2 to 3 times a week
  ☐ More than 3 times a week
  ☐ I don't drink
2–4 Teacher to check

Chapter 16

Exercise 16.01

1 a $T = 5$ **b** $D = 240$
2 a 180 km   **b** 4 h
3 80 km/h   **4** 275 km   **5** 8 km
6 a 0.15   **b** 9 min
7 7.5 min
8 a 2000 m = 2 km
  b 30 is minutes not hours.
  c 4 km/h
9 a 279 m   **b** 4.5 s
10 a 5400 m   **b** $11\frac{1}{9}$ s
11 a 750 km   **b** 10 h
  c 76 km/h   **d** 5 a.m. Tuesday

12 a 16.7 m/s, 25 m/s
13 40 320 km/h
14 Teacher to check. The distance will be in the order of 100 km.
15 a 8 km   **b** 4 km   **c** 1 hour
  d 4 km/h   **e** 2 hours   **f** 6 km/h
  g The second hour is uphill. In the first hour, his speed was 8 km/h and in the second hour his speed was only 4 km/h.
16 a i 2 p.m.   **ii** 11 a.m.
  b 12:45 p.m.   **c** 48 km   **d** 4.5 h
  e i 16 km/h **ii** 6.4 km/h **iii** 9.6 km/h
  iv 9.6 km/h
17 a 240 km
  b Sam, steeper graph, ending journey earlier.
  c 120 km   **d** 180 km
  e 10 a.m., 160 km
  f after 9:15 a.m., graph is steeper
  g 40 km/h   **h** B
18

Graph	Person	Distance from school
A	Luke	10 km
B	Peta	8 km
C	Wayne	4 km
D	Shelby	2 km

19 Teacher to check.

Exercise 16.02

1 a NW   **b** S
  c N or NW   **d** SE
2 a SW   **b** Poland, West Germany, UK
3 a NW   **b** SW   **c** S
  d W   **e** E   **f** NE
  g SE   **h** N   **i** E
4 A
5 a Condobolin   **b** Molong
  c Grenfell
6 a rain   **b** hot, dry   **c** rain
  d cold   **e** hot, dry   **f** rain
  g rain   **h** hot, dry

Exercise 16.03

1 a Helensvale Primary

b churches

c Helensvale Plaza

d

2 a H14   **b** A6   **c** J9 or J10

3 a West   **b** North   **c** South   **d** East

4 a 250 m   **b** 3.5 km   **c** 2.5 km   **d** 3 km

5 Doyalson Place

6 Walk north along Discovery Drive for approximately 500 m. Turn right into Ashford Rd. Quirindi Ct is the third street on the right.

7 Turn right at the next set of traffic lights into Discovery Drive. Follow Discovery Drive for approximately 3 km until you get to the roundabout at Helensvale High School. Turn left into Helensvale Rd. Follow Helensvale Rd to the motorway on ramp.

8 Drive north on Kingston and take the first right into Habana St. Take the first left into Milaroo Drive. Follow Milaroo Drive to the intersection with the Gold Coast Highway.

9 a East   **b** 450 m

10 a Walk north-west along Port Hacking Rd, then follow the road around to the right. Turn down Blamey Ave, the first on the left. Keep walking until you reach Mirral Rd.

b 500 m   **c** $7\frac{1}{2}$ minutes

Exercise 16.04

1 a 30 m   **b** 116 m   **c** 190 m

2 a 1 : 10 000   **b** 1 km   **c** 680 m

d 1.4 km   **e** 11:28 a.m.

3 a 800 m   **b** Approximately 13 minutes.

4 a 200 km   **b** 36 km/h

c The average speed is low. The road is probably not sealed and rough in places.

5 a 20 m   **b** 60 m and 115 m

c $4900

6 a Even numbers from 156 to 190 and 175, 177, 181.

b 129 m

c 311 m

d around 21 m

e It will flood during 1-in-50 floods.

7 a 11.6 km   **b** 2 h 54 min

c South west of North Era

Exercise 16.05

1 a *ABCD*, *AFD*, *AED*

b 27 km, 25 km, 24 km

c *AED*

2 a *A* to *B* via the curve is 17 km, *ADB* 16 km, *ACB* 17 km

b 15 km   **c** *ABCD*

3 a *ADC*, *ABC*, *ABDC*, *ADBC*

b *ADBC*

4 a via the hotel   **b** 6 km   **c** 7 km

5 The shortest routes involve going directly to Southside, then taking the appropriate road to the fire.

6 Head north-west along East Esplanade for 200 m and turn right into Belgrave St. Drive for 400 m and turn left onto Raglan St. Manly Oval is on your left when you turn in.

Exercise 16.06

1 155 km   **2** 120 km   **3** 98 km

4 a 3664   **b** 64 km/h

5 a 80 km   **b** 1 h 4 min   **c** 1:40 p.m.

d i 141 km/h

ii His average speed was 141 km/h, but at some times he was travelling faster and other times slower.

6 a 456 km   **b** 13 h 6 min

7 a Mittagong   **b** 170 km

c 1 h 53 min   **d** 37 km/h

8 $182

9 a 2:30 p.m.   **b** $18 each   **c** $11 each

10 10:59 a.m.

Exercise 16.07

1 a 7:30 a.m.   **b** 45 min   **c** 6 hours

d 93 km/h   **e** 3:17 a.m.   **f** 18 min

g 10:20 a.m.   **h** 26 hours 50 min

i $861

j $11, Many answers possible, for example, they wanted to see more of the countryside; or, the airline might charge more to include luggage; or, they didn't want a long stopover.

2 $170

3 a northwest   **b** 75 km/h

4 a $210   **b** $20

   c They need 2 caravans, but only 1 cabin.

   d No, teacher to discuss.

5 a Moockra Tower Lookout

   b Bruce

   c Middle Gorge

   d Teacher to check.

6 a 290 km   **b** 5 h 16 min

7 a They only need 1 room, but they would need 2 rooms at the Greenwood Lodge.

   b $381

8 a 0900. The 1345 flight lands at 1545, which only leaves 75 minutes for any delays and checking in for their flight. It may not be sufficient time and they could miss their flight home.

   b 2010 or 8:10 p.m.

Keyword activity

Across

5 coordinates   **8** distance

9 compass

Down

1 second   **2** kilometres

3 south   **4** scale

6 northwest   **7** speed

10 map

Chapter review 16

1 3 hours   **2** 15 km

3 20 m/s   **4** 25 m/s

5 4 km/h

6 a Average speed = total distance travelled ÷ (total time taken)

   b 52 km   **c** 10.4 km/h

7 Studio Village

8 1 km

9 At the end of Cannington Pl, turn left into Lindfield Rd. Take the second street on the right (Discovery Drive), then first left and second right.

10 a northwest   **b** Approximately 1400 m

   c Approximately 1 hour

   d Approximately 11:40 a.m.

11 a 15 (*ABDC*)   **b** 17 (*ABC*)

12 18 (*ABEDC*)

13 a 4WD (4-wheel-drive) vehicle

   b Alice Springs – Eridunda – Mt Ebenezer – Luntja Rd

   c Head south on Luntja Rd, turn right at Lasseter Highway.

   d 200 km   **e** Curtin Springs   **f** west

   g Northeast

Practice set 4

Section 1

1 C   **2** A   **3** D   **4** A

5 C   **6** B   **7** D   **8** B

Section 2

1 156 beats/min

2 $1826

3 Teacher to check

4 a Pacific Pines High School   **b** J9, 10

5 a 3.85 kWh   **b** $1.31

6 a 75 m/min   **b** 4.5 km/h

7 Teacher to check

8 a $8984   **b** $9883

   c $824   **d** Teacher to check

9 a SE   **b** N   **c** NW

10 Teacher to check

11 a 1405 kJ   **b** 42.15 minutes

12 a $17 384.64   **b** $16 640.25

   c Refund, $744.39

GLOSSARY AND INDEX

12-hour time: a.m. or p.m. time, for example, 2:35 p.m. (p. 269)

24-hour time: Time expressed as a 4-digit number with the first two digits representing the hour (from 00 to 23) and the last two digits representing the minutes, for example, 1435 representing 2:35 p.m. (p. 269)

allowable tax deduction: A part of a person's yearly income that is not taxed, such as work-related expenses or donations to charities. All deductions are subtracted from yearly income to determine **taxable income.** (p. 334).

allowance, worker's: Money paid to a worker for job-related expenses (for example, travel, special clothing, working in isolated or dangerous areas). (p. 244)

annual: Per year. (p. 160)

annual leave loading: Extra pay given to a worker during annual leave, usually 17.5% of 4 weeks' pay. (p. 247)

Australian Central Standard Time (ACST): The timezone for central Australia: the Northern Territory and South Australia. (p. 275)

Australian Eastern Standard Time (AEST): The timezone for eastern Australia: Queensland, NSW, the ACT, Victoria and Tasmania. (p. 275)

Australian Western Standard Time (AEST): The timezone for Western Australia. (p. 275)

bias: In statistics, an unwanted influence that stops a sample or survey from being representative of a population. (p. 352)

BIDMAS: *See* **order of operations**.

bonus: Extra pay for doing good work, meeting targets or deadlines. (p. 244)

break-even point: The point or value of sales at which a business stops making a loss and starts making a profit. (p. 210)

budget: A plan for managing money. (p. 340)

calories (Cal): An imperial (before metric) measure of the energy contained in food. (p. 312)

categorical data: Information represented as a category rather than a number, for example the makes of cars or the colours of eyes. (p. 178) Differs from **numerical data**.

census: Collection of information about every item or person of a population. (pp. 294, 296)

climate graph: A graph that shows the 12-month pattern of rainfall and temperature of a city by showing a column graph of rainfall and line graphs of maximum and minimum temperatures. (p. 105)

clustered column graph: A column graph with two or more columns stuck together for each category on the horizontal axis. (p. 94)

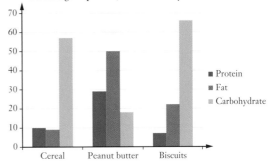

column graph: A graph consisting of columns of equal width. (pp. 93, 180)

compass direction: A direction shown by points on a compass, such as north (N), south (S), east (E), west (W), north-east (NE), north-west (NW), south-east (SE), south-west (SW). (p. 376)

ISBN 9780170412650

commission: The earnings of a sales person or agent; usually a percentage of the value of items sold. (p. 249)

constant: A number on its own in an equation. For example, in the equation $y = 2x + 3$, the constant is 3. (p. 200)

conversion graph: A graph that is used to convert between different units, such as between metric and imperial units of measurement, or between currencies in foreign currency exchange. (p. 96)

coordinates: A pair of values and/or letters for describing a location on a map, for example, D4. (p. 381)

cost price: The price it costs a shop to buy a product from the factory or wholesaler. (p. 41)

data: Statistical information. (p. 178)

decrease: To make smaller. (p. 40)

discount: Money deducted from the usual marked price. (p. 38)

distance-time graph: A line graph that describes a journey, by comparing distance on the vertical axis with time on the horizontal axis. The slope or steepness of the graph measures speed. (p. 368)

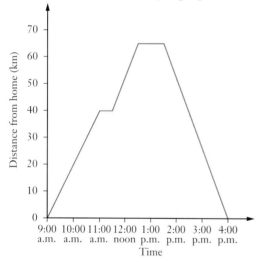

Monique's cycling trip

dot plot: A graph that uses dots to show frequencies of data scores. (p. 186). *See also* **stem-and-leaf plot**.

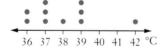

double time: A rate of overtime pay that is 2 times the normal hourly rate. (p. 241)

estimate: To make an educated guess for a numerical answer. (p. 15)

expense: The cost of spending. (p. 13)

fortnight: Two weeks. (p. 13)

frequency: The number of times a score or group of scores occurs in a data set. (p. 179)

frequency histogram: A column graph in which the height of each column represents the frequency of a single score or group of scores. There is no space between the columns. (p. 182)

frequency polygon: A line graph formed by joining the midpoints of the tops of the columns of a frequency histogram. (p. 110)

frequency table: A table that lists the frequency of each item in a data set. (p. 179)

fuel consumption: The rate at which a vehicle uses fuel, measured in litres/100 km. (p. 82)

gradient: The steepness of a line. (p. 199)

For the line graphed, gradient $= \dfrac{2}{1} = 2$

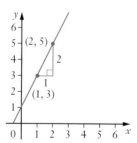

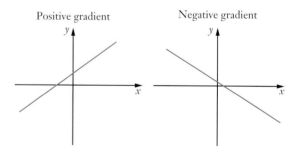

Positive gradient   Negative gradient

NELSON QMATHS 11. Essential Mathematics

ISBN 9780170412650

gross pay: Your pay before the tax is taken out. *See* also **net pay**.

GST: Goods and services tax. (p. 31)

heart rate: *See* **pulse**.

histogram: *See* **frequency histogram**.

holiday loading: *See* **annual leave loading**.

income: Money that is earned or gained (usually regularly). (p. 334)

income tax: A tax paid to the government based on the amount of income you earn. (p. 334)

increase: To make bigger. (p. 38)

integer: A positive or negative whole number or zero, for example, –3, 7, 0. (p. 8)

kilo-: One thousand.

kilogram: 1000 grams. (p. 32)

kilojoules (kJ): An measure of the energy contained in food. (p. 312)

kilometre: 1000 metres. (p. 20)

kilowatt (kW): A unit of electrical power equal to 1000 watts. (p. 320)

kilowatt hour (kWh): A unit of electrical energy equivalent to that used by one kilowatt of power in one hour. (p. 320)

line graph: A graph made of lines, often used to represent data that changes over time (p. 95)

loss: The amount of money lost by a shop on an item; loss = cost price – selling price. (p. 41)

map scale: *See* **scale**.

Medicare levy: A tax to cover the costs of the public health system, calculated as a percentage of a person's income. (p. 337)

net pay: Your pay after the tax has been taken out. *See* also **gross pay.**

numerical data: Data that involves numbers, such as heights or the number of cars owned. (p. 178)

order of operations: The order in which you calculate an expression with two or more operations: brackets, indices (powers), multiply and divide from left to right, add and subtract from left to right. This can be remembered using the initials BIDMAS. (p. 10)

outlier: An extreme (high or low) score in a data set that is very different from the other scores. (p. 190)

overtime: Woking beyond normal hours, paid at a higher rate. (p. 241)

per annum (p.a.): Per year. (p. 159)

PAYG tax: Pay-As-You-Go tax, paid every time you earn an income, an advance payment of your income tax due at the end of the financial year.

picture graph: A type of graph where pictures are used to indicate quantities. (p. 92)

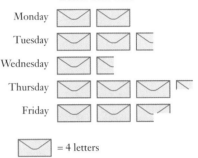

Letter deliveries

= 4 letters

pie graph (or **sector graph**): A graph using a circle divided into parts called sectors. (p. 99)

piecework: A type of work in which a worker is paid per item produced or processed, such as dressmaking or craftwork. (p. 249)

profit: The amount of money gained or made by a shop on an item; profit = selling price – cost price. (p. 41)

pulse (or **heart rate**): The rate at which a heart beats, measured in beats per minute. (p. 323)

quarterly: Every three months (quarter of a year). (p. 160)

random sample: A sample for which every member of a population has an equal chance of selection. (p. 298)

rate: A comparison of two quantities with different units, for example, km/h. (p. 72)

ratio: A comparison of two quantities with the same units, for example, 2 : 3. (p. 130)

retainer: A fixed amount of money paid to a salesperson that does not depend on sales, paid before commission is added. (p. 250)

royalty: A payment to an author, singer or artist for each copy of their work sold, usually a percentage of the total sales amount. (p. 249)

salary: Fixed earnings quoted as a yearly amount, but paid weekly, fortnightly or monthly. (p. 236) *See also* **wage**.

sample: A group of items selected from a population for surveying. (pp. 294, 296)

scale: (on a map or diagram) The ratio of scaled length to actual length, for example, a scale of 1 : 500 means that lengths represented on the map or diagram are actually 500 times larger in real life. (p. 383)

scale drawing: A diagram of an object, usually smaller, whose lengths are in the same ratio as the actual lengths of the object. (p. 144)

sector graph: *See* **pie graph**.

self-selected sample: A sample in which people volunteer to be part of the sample, such as an SMS poll or a website survey, so it is not really random. (p. 298)

selling price: The price a shop sells a product for. (p. 41)

simple interest: (or **flat rate interest**) Interest earned or charged only on the original amount of money (principal) invested or borrowed, different from compound interest. (p. 159)

solution: The detailed answer to a problem, or the method of solving it. (p. 8)

speed: A rate that compares distance travelled with time taken. Speed is often measured in kilometres per hour (km/h) or metres per second (m/s).

$$\text{speed} = \frac{\text{distance}}{\text{time}} \text{. (p. 366)}$$

stem-and-leaf plot: A 'number graph' that lists all the data scores, in groups. This stem-and-leaf plot shows 12 test scores, from 42 to 82. (p. 186) *See also* **dot plot**.

Stem	Leaf
4	2 5
5	0 2 8
6	6 7
7	3 5 7 7
8	2

step graph: A graph that has horizontal line segments, for example, parking station charges. (p. 98)

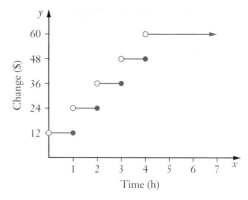

stratified sample: A sample consisting of a percentage of items from each 'strata' or 'layer' of a population. For example, a stratified sample from a population of 35% children and 65% adults should contain 35% children and 65% adults. (p. 298)

superannuation: A retirement fund that your employer pays into each time you are paid at work. It provides an income for you after you have completed your formal working life (p. 256)

systematic sample: A sample chosen by using a set pattern (for example, choosing every 10th number in a phone book). (p. 298)

tax debt: The amount by which the amount of **PAYG tax** already paid is below the amount of tax due. This is owed by the taxpayer to the Australian Tax Office (ATO). (p. 334)

tax deduction: *See* **allowable tax deduction**.

tax refund: The amount by which the amount of **PAYG tax** already paid is above the amount of tax due. This is given back to the taxpayer by the Australian Tax Office. (p. 334)

tax return: A form completed at the end of a financial year to state income earned, allowable deductions and tax already paid. Used to calculate a **tax refund** or **tax debt**. (p. 334).

taxable income: The part of a person's income that is taxed, equal to yearly income minus allowable deductions. (p. 334).

time-and-a-half: A rate of overtime pay that is 1.5 times the normal hourly rate. (p. 241)

timetable: A table displaying a schedule of events, for example, the times trains arrive. (p. 277)

timezone: A vertical zone where all locations have the same time of day, for example, central Australia (Northern Territory and South Australia) uses the Australian Central Standard Time zone (ACST). (p. 275)

unit price: In supermarket pricing, the price of one unit such as 100 g or 1 L. (p. 76)

vertical intercept: *See* **y-intercept**.

wage: An amount paid for work as a rate in dollars per hour. (p. 236)

y-intercept (or vertical intercept): The value at which a straight-line graph cuts the y-axis. For example, the y-intercept of this graph is 3. (p. 199)

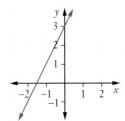

ISBN 9780170412650